Lean Behavior-Based Safety℠

BBS for Today's Realities

Shawn M. Galloway & Terry L. Mathis

Published by SCE Press

Cover design by Leigh Anne Griffin

ISBN 978-0-692-86826-3

Acknowledgements:

We would like to thank some of the people who helped us to make this book possible: our great consulting clients for whom we developed these methodologies, Traci Long and Lori Bowlin for their editing and formatting and unwavering commitment to our clients' success, other authorities in the areas we specialize in for their contributions, and our gracious wives and children who let us commit much of our time to complete the work we are so passionate about.

CONTENTS

FOREWORD

The methodologies in this book are the product of over 20 years of successfully implementing over 1,000 customized BBS processes in a wide variety of organizations and in over 40 countries. While the book is organized by topic, you will likely also sense a chronology of our evolving discoveries of better techniques and methods. This has truly been an example of continuous improvement that is still ongoing with each new project. The one constant in all this is the need for customization at each site. No two cultures are exactly alike and few sites have identical safety challenges which is why inflexible approaches to BBS have a significant failure rate. As workplace realities have evolved, so has our approach. Lean thinking and hard times have taught us great efficiencies that can be applied to BBS processes to make them faster, less expensive and more efficient. We sincerely hope these chapters expand your thinking about how you can customize a new or existing BBS process at your site.

Introduction

We realized nearly two decades ago what others are just discovering: that behavior-based safety (BBS) and Lean Six Sigma can be used together with good results. What others have yet to discover is that these two processes don't just complement each other in parallel, but dramatically improve each other when intertwined. Applying Lean principles to traditional BBS efforts makes it significantly easier and more efficient without compromising effectiveness. It also effectively re-energizes existing processes that are becoming routine.

Chances are you have a concept of what you think BBS is and how it works. We want to challenge that concept. Unlike many common approaches to BBS, Lean BBS® is an elegant tool to make safety better, quicker and more permanent, with lower costs and fewer worker hours. It is a way to identify and target transformational improvement opportunities a few at a time, identify any organizational factors that impact these opportunities, and align them in a way that will facilitate the desired changes well into the future. It can give your safety culture the ability to continuously improve. It can also adapt itself to your culture in innovative ways. If these features were not part of your initial concept of BBS, you are not alone. Most models of BBS attempt to do too much at once and die under their own weight.

Thirty years ago, behavior-based safety was touted by many to be the magic pill or silver bullet of safety. Today, it is often labeled as outdated or old thinking. We believe neither of these is true. BBS is simply one tool in a safety toolbox. It is not a holistic approach to safety, but neither is it irrelevant or outdated. An organization needs a way to direct, control and continuously improve discretionary worker behaviors, and BBS can be a highly-effective way to accomplish that. If you have not tried BBS, you have a great avenue to improved performance. If you have implemented BBS with less-than-stellar results, you have a new array of ways to make it work better.

Over these past thirty plus years, several specific processes for implementing BBS have struggled for marketplace dominance. Almost all of these approaches have been both inflexible and slow to evolve. Any methodology must eventually be enhanced or become obsolete. Moreover, any process that isn't efficient (i.e., Lean) in its search for increasing effectiveness will overburden those supporting it.

Lean BBS is not a singular approach, but a range of options that allow for maximum fit and customization. Each company, site or location will be different in many ways and can customize their process to fit their unique culture and situation while being true to the main principles necessary to success. A key theme throughout the book will be: "Do not attempt to make your organization fit a methodology; rather, make the methodology fit your organization."

There are three goals to this book:

1) Provide an understanding of the basics of BBS

2) Help the reader discover how to make a customized process work at their site

3) Discover opportunities to make an existing process more efficient and effective

Each chapter will work towards accomplishing each of the three goals. We recognize the scope of this book is not adequate to discuss all possible contingencies, nor provide a detailed step-by step how-to guide for full implementation of a process. But it can open up a world of new possibilities for those who have a singular view of what BBS is. It can also help to avoid taking an inflexible approach, which is the most common cause of failure in BBS processes. If you haven't discovered in your professional experiences so far, projects don't fail in the end, they fail in the beginning; and false starts almost always create barriers to future attempts. Read this book thoroughly and, with each chapter, consider which approach will work best at your site and in your culture. BBS, when used correctly and situationally, can be a great addition to your ever-growing safety excellence toolbox. We sincerely wish you great success.

Shawn M. Galloway

Terry L. Mathis

Part 1: Understanding and Misunderstanding Behavior-Based Safety

Chapter 1
What Is and Isn't Behavior-Based Safety

Not Everything Called BBS Really Is

As BBS became popular and accepted, many consultants and companies took the opportunity to capitalize on its success. Almost any safety process that impacted worker behavior in any way was labeled as BBS. Audit processes added a few human elements and called themselves BBS. Incentive programs changed their criteria from results to behaviors and called themselves BBS. People who participated in a BBS process at their place of employment resigned and declared themselves BBS consultants. They attempted to duplicate their employer's process at their new client's sites. The mainstream of what was originally BBS got hopelessly diluted.

Background of BBS and Understanding the Roles of Behavior in Safety

Human behaviors have always had a role in safety, and they always will. Behaviors were the primary, and sometimes only, tools for survival and remains today as the last tool when all else fails. When in an environment you do not control, when you lack the right tools or as systems fail, it is up to the individual to behave in a manner for self-preservation. This is popularized with the common statement, "You are the one responsible for your safety." This is not ideal; it is, however, reality.

Most of the modern world now places a priority first on conditional safety, with government regulation mandating that leadership provide a safe working environment. The acceptance of the safety hierarchy of controls also placed conditional safety as the first approach and behavioral controls as secondary or last. However, investment in conditional safety eventually reaches a point of diminished return, and when this occurs leaders often turn to behaviors as another approach to further safety improvement. Leaders of the organizations with the most excellent safety performance address both workplace conditions and worker behaviors, and continuously improve both, remaining aware that neither exists in a vacuum and either can impact the other.

Controlling vs. Influencing Behavior

Some behaviors in safety are mandatory and must be controlled as a part of leadership responsibility to maintain compliance. Others, however, can only be influenced. The tools to successfully manage, control and influence should not overlap. Unfortunately, they too often do, creating much controversy, fear and resistance from workers and from organized labor. Some approaches to BBS encourage this overlap and incorporate it in the methodology. Lean BBS recommends a careful separation of mandatory behaviors such as rules and procedures from discretionary behaviors such as ergonomics and body mechanics. Keeping rules and enforcement carefully separated from the behaviors BBS focuses on allows the process to be voluntary which can result in increased worker engagement.

Safety behaviors fall into two other important categories: injury/incident prevention behaviors and desirable safety culture behaviors. Within each of these categories, there are the two previously-mentioned types of behaviors: mandatory and discretionary (see Figure 1). It is vital to acknowledge this and ensure change tools are focused appropriately.

Figure 1: The Four Types of Safety Behaviors

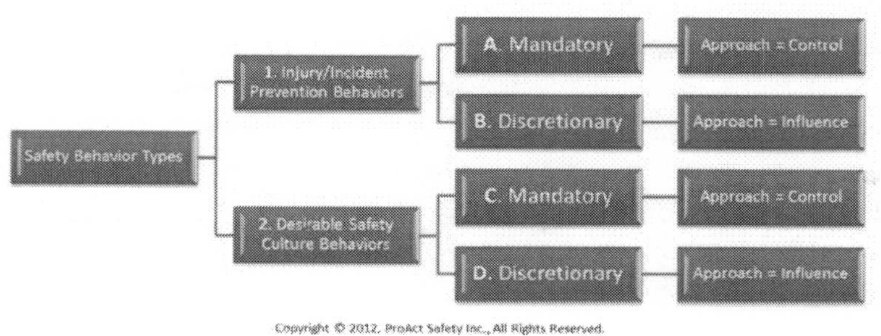

1.) Injury/Incident Prevention Behaviors

To prevent injuries, there are mandatory injury/incident prevention behaviors expected of individuals. In mature safety systems, these mandatory behaviors are covered by rules, policies, procedures, personal protective equipment, etc. Consistently enforcing these types of behaviors and controlling deviation are primarily the responsibility of management and supervision. Failing to perform these behaviors will result in disciplinary steps by many companies and government agencies. When it is stated that "Safety is a condition of employment," these are the behaviors being referred to and should be controlled.

Further, there are discretionary injury/incident prevention behaviors that generally go beyond compliance considerations. Keeping your eyes focused on your direction of travel and your body parts out of the path of potential moving parts are, for the most part, considered discretionary. It would be difficult to enforce these behaviors as rules. These types of behaviors are the primary focus of many behavior-based safety approaches. BBS is most often an employee-led approach and can be a great, situationally-appropriate tool. The Lean BBS approach is designed to focus on just one of the four types of behaviors in safety: discretionary injury/incident-prevention. Significant problems arise when BBS is directed at other behaviors, especially mandatory ones. BBS is most effective as a tool of influence, not another opportunity to control behavior mandatory to remain employed. This is especially important when peers are used as observers. A worker influencing another worker is different than the influence of supervisors. When mandatory behaviors are on BBS checklists, workers are often put in a conflicting position of acting as the supervisor rather than the fellow worker.

2.) Desirable Safety Culture Behaviors

All groups of individuals working together over an extended period of time create shared values and beliefs that eventually align safety behaviors within their culture. Safety cultures are nothing new; they have always been part of the organization. And, yes, you already have one. But is it the one you want? Is it providing value and a competitive advantage in your company and industry?

When an organization determines the elements, characteristics and capabilities of their desirable safety culture, the remaining two types of safety behaviors become increasingly visible. Individuals helping to achieve the ideal culture will comply with mandatory behaviors required of the safety culture (injury reporting, attending safety meetings, stopping the job for a safety concern, etc.) and discretionary cultural behaviors that exceed what is expected within the group (discussing safety issues with fellow employees, volunteering, identifying improvement opportunities, mentoring a new employee, etc.). The observation process can provide a model of how a worker should discuss safety issues with another worker. The list of other desirable cultural behaviors will differ for each organization depending on maturity and the role safety plays within organizational values and priorities.

Evolution

Those responsible for safety performance and culture must recognize a clear distinction between mandatory and discretionary behaviors and the roles they play in injury/incident prevention and culture. Tools to control behavior should be used differently than tools to influence behavior. Certainly control is a type of influence, but what occurs when the controller is not around? Forced change is often temporary. Influenced change is more likely to be permanent.

There are behaviors we expect in injury/incident prevention and culture which should be consistently enforced. However, excellence is not simply meeting the expected standards; it is exceeding them. To encourage excellence, it is necessary to have steps workers can take that are above and beyond the required behaviors. When taking steps towards excellence becomes the norm within the workforce, then excellence becomes cultural.

Goals of BBS

Organizations often have different reasons for adopting BBS other than simple accident reduction. Below is a list of some of the additional things a good BBS process can help to accomplish:

- Increase worker involvement in safety
- Focus workers on the highest-impact behaviors for accident reduction
- Make interpersonal safety communication (feedback & coaching) a cultural norm
- Organize workers to attack accidents from a different front
- Teach managers and supervisors to empower workers and support their safety efforts
- Introduce a proactive safety metric to enable cultural safety improvement
- Address behaviors not addressed by rules and procedures that can impact accidents
- Solicit discretionary worker effort to improve safety (people support what they help create)
- Support and supplement traditional safety efforts
- Identify and control barriers to safety (conditions that limit or restrict precautions)
- More accurately prioritize hazardous conditions by how they impact worker behaviors
- Drive safety efforts and responsibility down to the worker level

- Demystify safety through the concept of precautions and low-probability risks
- Develop a personal metric for safety more meaningful than TRIR/TIR
- Get workers thinking one step ahead of accidents
- Make the most important safety behaviors habitual
- Lower cultural tolerance for risk taking
- Get safety culture into a continuous-improvement loop
- Align every element of the safety culture in a safety strategy
- Make compliance a minimum standard for safety
- Measure common practice and the elements which influence it
- Give management a formula for safety culture development
- Form the culture around a core of RESULTS
- Enable the culture to better discover and share best practices
- Foster common proactive safety terminology
- Provide a quick-start guide to new employees and contractors

Customizing without Compromising

We have already stated Lean BBS is not a singular methodology, but a range of choices. When choosing among these options, it is critical to be true to the underlying principles of the process. There are four principles upon which Lean BBS is based, and keeping them in mind will help keep your process on course. You can accomplish these four principles in a wide range of ways, but make sure you have addressed all four. We have made an acronym to help you remember them as you progress through your choices: FILM.

The FILM for a Cultural Snapshot

A famous philanthropist once told Gandhi that his goal was to help mankind. Realizing the impossible enormity of this goal, Gandhi asked, "Which one?" We have to approach safety the same way when our clients tell us their goal is to reduce accidents. "Improving safety" entails so much; it is important to determine where to focus first.

Hundreds of organizations and thousands of sites worldwide have benefitted by using this simple four-part model to approach accident reduction. The model will help you be true to the basic principles of Lean BBS while you customize your approach. Additionally, it will help you create a leading indicator of safety that will more proactively measure and manage results. We refer to this model as the FILM to take snapshots of your safety culture and common practice.

FOCUS: If you asked your employees to focus on one thing to improve safety, what would it be? Would it improve safety and change the results you are currently getting? If your focus in safety hasn't changed performance, you might not be focusing on the right things. Interview groups of employees and ask them, "What is your greatest risk on the job?" or, "What risk do you spend the most energy trying to avoid?" If you get too many answers, worker safety efforts are not focused. Being focused does not mean you quit looking at the big picture of safety; it means you also concentrate on a specific area to improve.

Unfortunately, two problems are very common when developing a correct safety focus: kaizen (continuous small improvements) and the standard classifications of accidents. Most organizations should not work on one small improvement, but a huge one. This school of thought is called "transformational thinking." The challenge is to discover the one thing that would improve safety most. However, "overcoming hand injuries" may still be too broad a goal. You might need to focus on "pinch points" or "hand-tool selection" or other, more specific issues that impact hand injuries. Rather than just trying to improve safety in general, work on truly conquering a specific safety issue.

The enactment of OSHA in 1970 has done a great deal to decrease incident rates. Now, however, many US organizations have reached plateaus that represent a low level of accidents. Interestingly, most US companies are putting forth the right amount of energy into safety. The next step in safety performance improvement will not come from additional energy; it will come from a continuous focus on the transformational. Throwing resources at a problem was an accepted practice when the average company had dispensable resources available to them. Today, it has become critical to ensure improvement efforts are focused on value-add and efficiency. These criteria will also ensure the results can be sustainable.

A Lean BBS focus is most often between two and six behavioral precautions. Even if you ultimately need to address more behaviors, it is best to do so a few at a time. Traditional BBS checklists have too many behaviors to create true focus. If workers cannot remember and recite the targeted precautions, they will never become habitual.

INFLUENCE: People do things the way they do for a reason. Change the reason, and you can change how people do their jobs.

The late quality gurus, Dr. W. Edwards Deming and Joseph M. Juran, dispelled the myth years ago that all problems would be solved if people just did their jobs well. People tend to do their jobs well when they are adequately trained, appropriately supervised, and have ergonomically engineered job stations and processes. Personal mistakes and deliberate risks are a part of safety also, but have outside influences.

Organizations that align influences toward safety have better results than those who simply try to enforce compliance with safety standards. Kerry Patterson and his associates did a remarkable job shedding light on this topic in their 2007 book, *Influencer*. They pointed out that changing influences was not only the best way to change behaviors, but also produced the most permanent change.

In Lean BBS, observers must identify what is influencing safety-related behaviors by using the next principle.

LISTEN: I recently asked a supervisor why his workers performed a task a certain way. He responded he didn't know and wasn't sure how to find out. I shared with him a time-tested process I have used many times: I walked up to the employee and asked him why he did the task that way.

The worker spent the next three minutes giving me an insightful and thorough explanation that revealed a serious barrier to doing the job a safer way. The principle here is "people often know why they do what they do," and you can know too if you ask and listen.

Leading and directing employees are important skills for management and supervision. Listening is an equally-important part that is often ignored. A true understanding of common practice and safety culture isn't simply discovering what people do, but why they do it.

The "why" part of this equation is where the true power to change things lies. When you get an employee to be safe in spite of the influences to take risks, you make that employee safer. When you change the influences to take risks, you fix the problem for every employee who will do that job in the future. Some influences cannot be changed and must be addressed through difficult behavioral strategies. Identifying and addressing these can also leave a legacy of safety for future workers.

MEASURE: Recent articles and books have criticized safety measurements and how lagging indicators are inadequate for further improvement. The FILM model allows you to measure your progress toward a specific safety goal and to see how improving this goal impacts the lagging indicators of safety (in the targeted category).

The metric is simply how much you have improved performance toward a target. In a Lean BBS process, this is the percent safe of the behavior impacting the specific type of accident. As you improve your efforts, you can see the impact on your accident rates, severity, costs, etc. If you focus on a certain type of injury, this is where you should see results. The leading indicator tells you if you are working your plan, and the lagging indicator tells you if your plan is working to reduce your accidents. The multiple indicators allow for more effective diagnostics and faster corrections to safety improvement strategies.

Dictionary.com defines *respond* as "to react favorably," whereas *react* is "to act in opposition, as against some force." To be effectively proactive in safety, an organization must *respond* to insight rather than simply reacting to an unplanned, undesired event or outcome. To achieve and sustain safety excellence, energy will need to be focused; influences must be understood and responded to by listening to the rationale for common practices. Listening to find out what influences decisions, managing those influences to support the desired behavior, and measuring how effectively workers change targeted behaviors will change the way we measure safety performance.

If this model sounds too simple to work, remember this is not a theory nor a proposal, but a proven model being utilized by organizations worldwide. The remaining challenges are in the details of how to best make it fit your site and make it sustainable within your culture.

If your organization feels it cannot do all four of these things at once, start them one at a time. First, focus your workforce on one specific safety issue. Then align the known influences that would support workers' efforts. As you progress, begin to ask what other influences impact the targeted improvements. Lastly, begin to measure the percentage of time workers follow the targeted improvement strategy and how many times they do not. Each step can add value and impact safety performance alone. However, once all four are in place, the results are even greater than the sum of each step alone.

The Cliff Analogy

There are two elements to almost all accidents: conditions and behaviors. The condition in this analogy is a cliff, something you can fall off of. If a person walks to the edge of the cliff, he is performing an at-risk behavior. Not every person who walks to the edge of the cliff will fall off, but every person who falls walked to the edge.

If you have people working in hazardous conditions, there are many ways you can prevent them from getting injured. One way is to use conditional safety by working with the cliff itself to try to make the workplace safer for the worker. This can be done by adding barrier walls, for example. Another way is to use intermediate interventions such as personal-protective equipment (PPE). It won't prevent an accident, but will lessen the resulting severity if an accident occurs.

Regardless of all the efforts that can be made to make the cliff safer using traditional safety, it is not possible in all instances to make conditions 100% safe. There will always be a failure rate. When a failure occurs, you try to learn from it so the same mistakes aren't repeated. This is determined through an accident investigation. This is the *reactive* part of safety. After an accident has happened, how can we understand it better, study it thoroughly, and prevent it from happening again?

Behavior-based safety is a proactive, rather than reactive, intervention. Rather than waiting until someone falls off the cliff, we identify a behavior (such as keeping a 10-foot distance between you and the edge), and try to instill that behavior in the organization.

Instead of measuring data *after* an accident, you can measure before accidents occur. How many people walk out to the edge of the cliff? How many people keep a 10-foot distance? The ratio of people who stay 10 feet from the edge gives you a leading indicator to measure: percent safe.

No matter how good you are at conditional safety, you will always have to address the behavioral element of safety if you really want to achieve safety excellence. In traditional safety, what usually gets addressed are the mandatory behaviors covered by rules and procedures. If these alone don't eliminate all accidents, it's time to address those discretionary behaviors, which is what BBS does best.

Accident rates tell us mandatory safety is not enough. In other words, it's driven by its own failure rate. The problem with that is, as you get better and better at safety, the accident data starts to go away also. But, before it disappears, it loses statistical significance. It becomes random data that no longer has addressable trends.

If accident data is all you have to work with, you end up avoiding failure rather than achieving success. You don't know what safety success looks like except the absence of accidents. So it's a negative definition. And you drive this negative definition into the minds of the workers. This definition drives workers to think, "If I don't have an accident, then I must be performing my work safely." It is easy and, unfortunately, common for people to think this way.

Think about all of the safety pictures of risky situations you have received via email or have seen posted online. What was the most likely outcome of one of those dangerous situations? Probably nothing. Guess why there are so many photos of risks and so few of accidents? Because in most cases the worker went home and the company called it another "safe day" and changed the sign to add one more day since a recordable or a lost time. If this is the only measurement you have in safety, you may be sending the wrong message.

In any case, you are ignoring the presence of low-probability risks. These are risks that don't result in an accident every time you take them. In fact, these types of risks only produce an accident one time in hundreds or thousands of instances. It might take hundreds of times for people to walk out to the edge of the cliff before one falls off. However, the only effective way to prevent the fall is to reduce the risk pool of people walking out the edge. Remember the old saying, "It is better to be careful a thousand times than to be dead once!"

There are two ways to not have an accident. One is being safe by minimizing risk and taking the right precautions, and the other is simply by being lucky. With things that have a low probability of turning into an accident, you're going to be lucky more often than not.

Another issue is that low-probability risks fly under the radar of common sense and experience so, unfortunately, a lot of risky behavior becomes self-reinforcing. When a worker takes a risk, regardless of the reason, and doesn't have an accident, the lack of a negative consequence tends to reinforce the risk taking. Even when the risk does result in an accident, if the injury is minor the worker doesn't necessarily change behavior. If the resulting injury is more severe it can change the behavior, but the accident has already taken place and the change is reactive rather than proactive. The goal of BBS is to change behaviors BEFORE they turn into accidents, not just react to them after a worker is injured.

Every time an accident happens, you try to backtrack into why it happened and determine where your system fell apart. Was this the workers' fault? Was it a conditional fault? Was it something we didn't think of? Do we not have a rule we ought to have? You are managing by exception. It's much more effective to identify the precautions that could potentially control the risks and try to identify the control points of accidents. Interestingly, the control points of accidents are most often behavioral.

Sometimes statements like, "All accidents are preventable," tend to cause conflict. What the worker thinks is, "Once I get out to the edge of that cliff and a gust of wind catches me, and I'm starting to fall off of the cliff, I can't prevent it. There is no prevention." Well certainly the worker has a legitimate point. There is a point of no return in every cycle leading up to an accident. You can go so far there is no prevention beyond that point. This fact makes it even more critical to manage safety proactively through control points.

The prevention point is usually a step or two before the worker begins to sense the risk. Workers need to stop thinking "safety is not having an accident" or "safety is not falling off the cliff," but instead need to think "safety is keeping a ten foot distance from the edge of the cliff." Because if you're ten feet back from the edge of the cliff, the gust of wind won't push you off. You have reaction time and space between you and the danger. So can you keep a buffer zone? We already think that way in driving. How closely are you following the car ahead of you? If they slammed on their brakes, would you have stopping time? A lot of times we don't think that way in industrial settings.

Traditional safety has some basic weaknesses. It is driven, as mentioned earlier, by lagging indicators: accident rates, accident data and the cost of accidents. It also relies on a reporting system. Generally, the only way you know there was an accident or a near miss is if someone reports it. You have probably discovered that reporting systems are imperfect. It doesn't take much of an influence in your organization to suppress reporting or drive it completely underground.

Identify how safety is defined in your organization. Do people perceive and appreciate the risks in their job? This is only part of the equation, for if one person is influenced to take a risk, usually the influence will impact others as well. Don't let your root-cause analysis stop when you can't answer the "why" question at the behavioral level. While this certainly provides an opportunity to identify a prevention or control point for an individual, let's get past the "who did what" in safety and keep working to understand the organizational influences: obstacles or barriers that encourage risk at the cultural level. Don't mistake common practice for culture. The "way we do things around here" is common practice. The underlying agreements and influences that shape the set of behaviors is the culture.

BBS and Traditional Safety

Although some have tried to use BBS to replace or be redundant to traditional safety, that is not its best use. Traditional safety addresses mandatory behaviors of safety through rules, procedures, required PPE, etc. If traditional safety is not totally effective, it should be improved, not replaced or repeated. BBS works best as a supplemental process to go above and beyond traditional safety by addressing the discretionary behaviors critical to safety not addressed with traditional efforts. This figure illustrates how this can be structured.

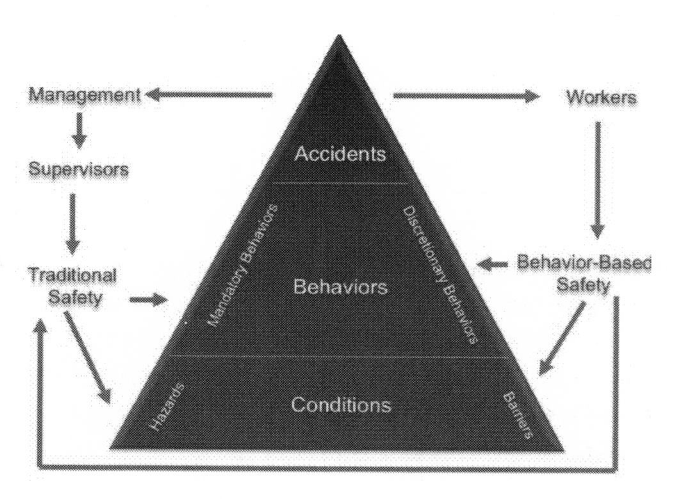

In traditional safety, managers review accident data and share their analysis with supervisors and through organizational processes such as safety meetings and training. They address the mandatory behaviors and hazardous conditions they feel can potentially prevent similar future accidents.

In BBS, a steering team of workers analyzes accident data and develops a checklist and observation process to address the discretionary behaviors and the barriers that get in the way of safety.

When you combine traditional safety and BBS, you are attacking accidents from two different fronts. This combined approach, with cooperation between the two efforts, has proven highly effective at efficiently eliminating accidents.

Another way to think about this is to envision the limitations of traditional safety as a brick wall. Where traditional safety maxes out, BBS picks up and moves safety to a new level not possible with traditional safety efforts alone.

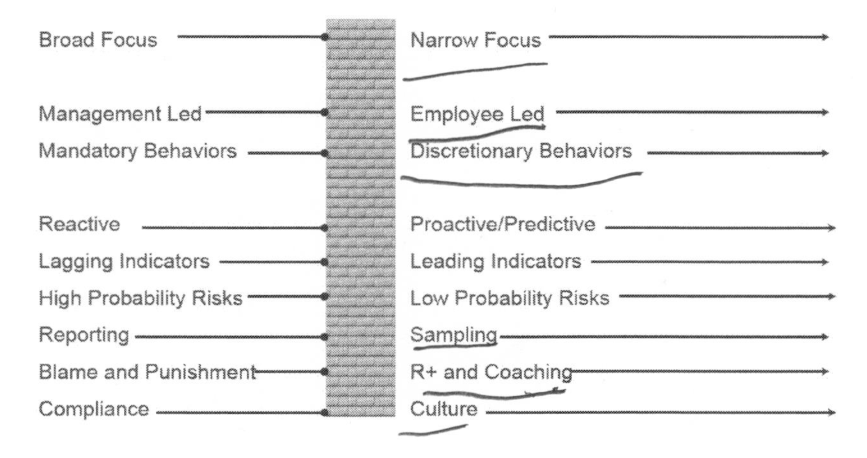

Existing BBS Processes

If you already have a BBS process, can you rethink the relationship between your traditional safety efforts and BBS? Do the two complement each other or compete? Do your workers understand the basic concepts of low-probability risks and proactive versus reactive safety interventions? Do workers understand their role in safety is to comply with rules and procedures (traditional safety) but to also go above and beyond (BBS)? Could some realignment and retraining breathe new life into your existing process?

Chapter 2

Truths and Misconceptions about BBS

As mentioned previously, the popularity of BBS attracted many practitioners to re-label what they offered as BBS. The legitimate founders and practitioners of mainstream BBS disagree on some details of how best to practice it, but have developed a common core of practices on which they do agree. This chapter lays out what we believe are the most important aspects of BBS, how we think they should best be addressed and what should be avoided. Again, Lean BBS is a way of thinking and a range of options rather than a rigid and unalterable methodology.

We believe BBS is an effective, situationally-appropriate tool for addressing a certain type of safety behavior. When the limitations of control, systems and conditional efforts have been reached, BBS can provide substantial additional value. Recognizing how this tool compliments existing processes is just as important as recognizing what often leads to failure in both newer and mature processes.

Based on experiences implementing hundreds of customized approaches globally and improving as many existing processes, the common pitfalls to avoid are:

Forced Effort or Involvement - Behavioral safety processes that force involvement typically result in "voluntold" teams and less than desirable levels of engagement and safety improvement. As an example of forced effort, how much passion and willing discretional energy do you provide to paying your taxes every year? If a person is working toward an accomplishment because they have to, rather than want to, expect efforts to achieve minimal expectations rather than efforts that exceed them.

This is also true when organizations require a quota of observations to be accomplished by every employee. Individuals who do not want to be involved might not leave the observee with a positive experience, thus perpetuating further negativity towards the process. Remember this principle: Forced change is almost always temporary. When the force goes away, so does the change.

A "Gotcha" Approach - Several processes either do not announce or do not ask permission to conduct an observation. These processes have the misguided impression that spying is a better approach to see true common practice. While candid observations can provide accurate data on common practice, they also damage to relationships and culture. Clandestine Human Intelligence is often used within governments against other governments. It is not a tool to create cultures that break the "us versus them" mentality and work together to be excellent in any area of operational performance.

Effective processes either ask permission or, at minimum, announce the observation to make those to be observed aware. The goal is to see if workers are willing to take the targeted precautions and if doing so is possible. If it is, and we are certain these precautions represent safe practice, then encourage the person to continue taking the specific precautions. If they are not being taken, find out why.

Information Used for Discipline - Behavior-based safety processes that are effective collect insightful information into common practices of work activities. This insight helps determine how to help employees work as safely as possible. How is your data used? Some processes have resulted in employees receiving discipline for activities observed during behavioral observations. This is a gray area in many processes and requires careful consideration. You don't want to tolerate non-compliance but you also don't want your BBS process to be perceived as a safety policing process.

Organizations need safety rules and accountability for enforcing these rules. This responsibility, however, often lies with those in supervision and management. This is why many effective processes ask employees to conduct the behavioral observations while asking supervisors to play a very specific, customized support role based on the trust levels within the organizational culture. These effective, employee-led behavioral processes focus on discretionary behaviors and the management systems focus on the mandatory behaviors. (There are also opportunities to teach the observation and coaching skills to supervisors to become safety coaches. However, take care to ensure the processes are separated if trust issues exist.)

Asking employees to observe for, and collect data on, both mandatory and discretionary behaviors on a checklist is fraught with complications. While the responsibility of an employee witnessing another employee violating a safety rule doesn't go away, what are they to do if this is witnessed during an observation? Our most common recommendation is to tear up the card, stop the observation and intervene. Care is needed because if employees feel they will be disciplined following an observation they may not be willing to being observed or participate in other ways. If the process produces negative reactions and attitudes, how valuable is it? What is the perception of measurement in your organization? Is the goal of measurement to gather insight to learn, improve and remove all the barriers to safe performance? Or is the goal to catch people doing something wrong, hold them accountable and place blame? If metrics are viewed as tools for punishment and BBS produces another metric, expect negative reactions. Discipline is a necessary part of traditional safety, but can damage BBS. If possible, keep them carefully separated.

Lack of Action Plans or Visible Progress - Many processes focus on number of observations and tend to forget the goal of the process is improving safety. Behavior-based safety is not the silver bullet or magical solution some have made it out to be. It is simply another tool in the ever-improving safety toolbox. But this tool does serve a purpose: to gather insight into common practice and the reasons why precautions are not being taken or cannot be taken. This, then, requires the organization to act on addressing the reasons by creating action plans to either improve what is influencing workplace behaviors or improve conditions if the influences on risks are understood.

Consider your current safety committee. How many detailed successes have they had that you can recall from memory? The average BBS steering committee or team fails to celebrate and communicate their results. If the team is not creating data-driven action plans, completing the action plans and communicating their progress, will they be perceived as successful? Who wants to join teams viewed as ineffective? Work to create an environment where anyone approached can recall three successes of the behavior-based safety efforts. Consider expanding this to other teams in safety. If we want employees to volunteer their efforts to improve safety, they should feel they are joining a winning team, or at least one giving it their all.

Unfocused or Wrongly-Focused Efforts - Observations are often performed because there is a requirement to perform them. If people feel the only thing measured or reinforced is the number of observations, getting observations completed will not be your problem; safety improvement will be. If we recognize the goal of BBS is to improve safety, then why do we observe for the sake of observing? Why do we still blanket observations on any day of week, time of day, tenure of employee, tenure on task, trained versus not trained, icy versus humid, etc.? Consider targeting your observations where they are needed and will have the maximum impact.

Regretfully, many processes are still observing far too many behavioral precautions. A significant amount of brain science has proven that a focus beyond four to seven items is not a focus. If you have to rely on your checklist to "shape behaviors," you are using the most expensive and unsustainable resource available. Even if you have 15-25 behaviors that could improve safety, you will be more effective working on them 4-7 at a time than all at once.

Sustainability is an Afterthought - Far too many methodologies rely on implementation approaches that create a dependency on external consultants. If a company is to achieve sustainable performance in safety, then tools and capabilities critical to facilitate this need to be owned by the company. While this might not always be practical, it is realistically necessary.

As many organizations with mature BBS processes discovered, when they become unwilling or unable to make royalty payments for intellectual property, software subscriptions, or training materials fees to outside consulting companies, their BBS processes suffered. Use consultants if needed, but don't become dependent on them long-term. Work aggressively to ensure the tools necessary to achieve and sustain safety excellence are internal, not external. Seek out new ideas that can be retained and internalized into the structure to ensure evolution towards the desirable performance.

Expecting Miracles - The goal of a behavioral process is to provide simple strategies people can internalize quickly and to remove the obstacles and barriers to safe performance. Behavioral safety tools are primarily focused on workers who have experienced many things in life that provide a framework for their decision. We cannot expect to change that framework overnight.

Some BBS processes rely on confrontation to change worker behaviors. Simply tell the worker they are at risk and should do the job differently, and the problem is solved. Not so! Behaviors are chosen for a reason (whether knowingly or subconsciously). Finding the reasons why workers take risks or fail to take precautions and addressing these issues takes time. Even after we change underlying influences, the behaviors will take time to respond. New behaviors require additional time to become automatic or habitual. Expecting immediate change and impact on accidents is unrealistic. Criticizing the people involved in BBS for accidents in the early stages of the process has damaged and even destroyed BBS processes. Make sure everyone has realistic expectations and is informed of progress as the process begins and matures.

Stopping Rather Than Starting Behavior - Safety has too often been defined as what workers should *not* do rather than what they *should* do. Many BBS processes, likewise, create checklists of risks to "not take" rather than precautions to take. This approach is not impossible, but does tend to have negative side effects.

The tools behavioral sciences use to stop behaviors can be quite effective, but they tend to damage or destroy interpersonal relationships and cultures. These approaches provide negative consequences for taking risks. In order to be effective, they must be timely and consistent. Focusing on what workers do wrong rather than what they do right creates animosity rather than teamwork. Observers in such a process can be viewed as critics rather than friends.

The tools to start or promote behavior, on the other hand, can build strong relationships and cultures. They are supportive rather than critical. They coach people to be better rather than criticizing them into being less bad. They seek to achieve success rather than to simply avoid failure. They create a vision of how safety success looks rather than simply the lack of failure. Safety becomes the positive practice of taking precautions rather than the negative practice of avoiding risks. Observers in such processes tend to be viewed as friend rather than foe, coach rather than cop.

Stopping at Behavior - Behaviors are not the root cause of accidents or incidents. Finding a behavior that contributed to an accident and blaming whoever did it is not an effective corrective action. Targeting the behaviors that caused accidents in the past on a BBS checklist and simply telling workers not to take those risks is not an effective behavioral strategy.

There is a common quip among some safety professionals: "What is the root cause of all slips, trips and falls? Gravity!" All too often, incident investigations focus on employee behavior as the root cause and follow-up action plans such as: "Pay attention," "Employee needs to be aware," "Employee needs to not be distracted," or the all too often unrelated one, "Employee needs re-training."

People do things for a reason; however, many investigations end at employee behavior because the reason for the behavior is not determined, or is frankly not sought out. If your goal is to change or improve performance you must engage the individual in discussions to learn the reason(s) for the lack of performance. Very few employees go to work with the intentions of getting injured. Most do their job the way they know how, the way they normally do it. They do it that way because of influences that have shaped their practices over time. If you simply blame the worker or try to convince them to change, the influences that shaped their practices continue to impact them and anyone else that takes their place.

The most effective BBS processes do not simply try to confront workers and force them to change. They seek to find and address the influences that shape behaviors. Doing so not only impacts the one worker, but all who follow.

Process Orientation, Falling in Love with the Methodology - It is easy to become focused on a process when you practice it regularly. However, it is important to remember what the process is designed to accomplish. Many mature BBS processes stress the importance of hitting a target number of observations in order to be effective. Steering teams and leaders have become obsessed with the number of observations and forget the observations should accomplish behavioral change, and the behavioral change should accomplish improved safety performance. If the observations are taking place but are not effective, the process is not working and needs to be changed. Don't mistake activity for results!

Effective BBS leaders constantly compare their process metrics with results metrics and make sure their efforts are producing the desired results. If they are not, adjustments are made and monitored. The process gets regular reality checks. Don't let your leaders become process oriented at the cost of being results oriented.

BBS is Viewed as the Safety Excellence Strategy - Our 2013 book, *STEPS to Safety Culture Excellence*, points out the need for an overarching strategy for safety efforts. BBS can be a part of such a strategy, but cannot be and should not be viewed as the entire strategy. Strategy should define the vision of what safety excellence looks like in terms of management style, focus, culture and other critical elements. Safety programs such as BBS can certainly be utilized to accomplish one or more of those strategic goals, but it is not a replacement for the strategy.

In fact, programmatic thinking is almost the opposite of strategic thinking. Too many organizations see the need for better safety performance and simply try to use a safety program or set of programs to address it. If these programs are not aligned with strategic goals and objectives, they can actually work against each other. Be careful not to view BBS or any other safety program as a strategy!

Reasons for Union Resistance

If you research BBS on the internet, you'll find that some organized labor organizations have taken an opposing position to behavioral approaches to safety. Their early experiences with the first attempts to perfect this approach were unfortunate and damaging. However, most unions were formed with safety as one of their core values and improvement objectives. Many have embraced BBS as an ally in that objective and others have softened their resistance as BBS processes have evolved away from the methods that caused the initial resistance. We call these flawed methods the "seven deadly sins" and will now discuss how to avoid them in your BBS process.

Seven Deadly Sins of Behavior-Based Safety

Behavior-based safety has a proven track record of success. However, several unions have openly stated their opposition to BBS online.

We should recognize their concerns and complaints are real and valid. Some approaches to BBS have caused problems, and the results have been negative for unions and their members. But the next question should be: Is this problem the result of a flaw in the core philosophy of BBS or is it the result of poor methodology? The fact there have been hundreds of successful union-friendly BBS methods suggests the latter.

The truth is, BBS is a label applied to everything from safety incentive tokens to some very rigid and structured processes. Many of these processes have evolved over the years, and the consultants who designed them have changed their positions about some basic issues. Putting a single label on all these varied methods is misleading and inaccurate. However, if you subscribe to the idea that encouraging the use of specific precautions (safe behaviors) can decrease accidents, then the question becomes how to best do this and not cause negative impact like some approaches have created.

Each union position paper was written as a criticism of a particular approach to behavior-based safety; but, cumulatively, they form an overall critique of methods that are not only union unfriendly, but altogether ineffective. Our research cited 22 separate complaints we grouped into seven categories. Unfortunately, these seven methods are commonly used in many BBS approaches and almost guarantee union resistance as well as sub-optimized results.

THE SEVEN DEADLY SINS OF BBS ARE:

Blaming (believing, teaching or assuming that most accidents are caused by unsafe behaviors of workers) - Starting with this flawed premise creates a shaky foundation and instant animosity for a behavioral approach. Studies often cited to make this point are questionable and misquoted. Most classification of accidents into behavioral categories referred to prevention rather than root cause. One study stated if anyone could have done anything differently to prevent the accident, it was classified as caused by worker behavior.

Dean Gano, who developed a problem-solving methodology for NASA, argues in his book, *Apollo Root Cause Analysis*, that behaviors can never be the root cause of an accident since there is always a cause for the behavior. Starting BBS with such statements or assumptions suggests workers are to blame and must solve their own problems. Ignoring conditional and organizational issues that can cause both accidents and unsafe behaviors is a formula for failure: both to produce maximum results and to solicit union support.

Confronting (the belief BBS must target the unsafe behaviors that cause accidents and eliminate them by worker-to-worker confrontation) - The number one reluctance of workers to be observers is the fear of confrontation. They are willing to watch and identify potential risks, but they truly dread having to confront and convince their fellow workers to change.

There is a sense of pride in the way work is performed (especially among experienced workers), and such confrontations are a rude invasion of this pride. The first reaction is usually, "What makes you think you know more about safety than I do?" The training observers receive in most BBS processes falls short of qualifying them as safety experts. It falls completely short of preparing them to successfully confront and change behavior on the spot. The whole idea of confrontation assumes the problem can be solved by the individual and ignores the impact of other influences (i.e., conditional, organizational, cultural, etc.). Kerry Patterson, et al., in their recent book, *Influencer*, suggest that direct confrontation is almost always ineffective in producing a change in behavior.

Idealism (the belief BBS is a silver bullet to replace your other safety efforts) - Some academic experts in behavioral psychology have espoused the theory BBS is some kind of miracle cure for all that ails safety. Workers could potentially absolve managers of all responsibility in safety and give them a handy scapegoat for anything that goes wrong. In addition, managers may think they can save money in their budget. Instead of spending money to fix things and make them safer, they can simply alter the workers' behaviors to avoid the hazards. BBS has been most successful as a supplement to traditional safety efforts, not a replacement or redundant process.

Punishing (the belief it is okay to use punishment for failure to shape behaviors) - Discipline is a tool that infers blame and willful disobedience. Attempts to use discipline in voluntary processes almost always cause resistance and ultimate failure. When early behavior-based safety processes tried to use discipline as a tool to establish behavioral change, unions instantly protested.

This approach created the perception BBS was being used to get union brothers and sisters spying on each other. Including behaviors on a BBS checklist that overlap or duplicate safety rules or procedures almost ensures punishment will follow observations.

Isolationism (the belief management should be completely omitted from BBS processes) - Some approaches to BBS utilize workers exclusively and ask managers and supervisors to take a hands-off position. This hinders the BBS process from being able to address organizational issues and furthers the stigma of blaming workers and expecting them to work out their own behavioral problems.

Exclusion (the belief it is not necessary to involve unions in the decision to implement a behavioral approach) - Many unions are completely excluded from the decision to apply BBS and from any discussions about how to structure the process or select participants.

Unions care deeply about the safety of employees. Not involving the elected representatives in collaborative discussions about an employee-led safety process is an ineffective change strategy and, most importantly, disrespectful.

Inflexibility (the belief that one form or methodology of BBS is right for every site) - Ignoring cultural, regional, organizational and conditional differences from site-to-site and organization-to-organization was the norm among the early approaches to BBS. Academics tend to seek elegant, universal solutions and sometimes overlook the significance of site-to-site differences. Unions have good ideas of how to better fit safety processes to their sites and they were largely or completely ignored in favor of the "perfect" solution.

Unions have resisted behavior-based safety based on each of these seven problems, but BBS can be implemented in a union-friendly way that does not include these issues. In fact, unions embrace BBS when each of these seven concerns are carefully replaced with more effective and union-sensitive approaches.

The right approach includes the following:

1. Rather than fixing the blame, focus on fixing the problems.

2. Realize people make behavioral choices for a reason. If you don't change the reason, you probably won't change the behavior. So rather than confronting a fellow worker taking a risk, find out what is influencing that behavior, document it and take it to a steering team who will prioritize and address the issues.

3. Acknowledge BBS is no silver bullet, but another tool in your safety toolbox. Carefully separate BBS from traditional safety programs and allow them to work synergistically together without duplication or overlap.

4. Carefully separate any punishment from the process. BBS should be separate from traditional

safety, and behaviors on BBS checklists should not overlap with rules and procedures. This ensures no one is disciplined for BBS observation data.

5. Define management and supervisor roles, responsibilities and expectations in BBS in such a way that they support, without taking over the process. Enforce these guidelines to ensure they are executed properly.

6. Include unions in the decision to implement BBS, and in the design and customization of the process for the site. Their input is valuable and essential to success.

7. Stay true to the basic tenets of BBS, but customize and innovate approaches to fit the culture, site and any other programs in place, such as 5S, Lean or Six Sigma.

When practiced and implemented with sensitivity to union concerns, behavior-based safety can achieve superior results without encountering the resistance of other approaches. Not only does the adoption of these methods enable greater success, but it enhances the durability and fit of the BBS process to the site in a way that makes the process more sustainable long-term.

Top Union Concerns about Behavior-Based Safety Processes

Union officials have expressed the following concerns with utilizing BBS at their sites.

1. BBS is based on the belief almost all accidents result from unsafe acts. The union believes there are

typically multiple root causes, mainly environmental.

2. BBS enlists inspection "cops" to observe workers.

3. BBS blames accidents on the workers.

4. Use of discipline.

5. BBS tries to change worker behavior - and companies will focus less on workplace hazards. Tries to teach workers to work safely around a hazard instead of fixing it.

a. Occupational Safety & Health Act of 1970, General Duty Clause – 5(a)(1) Each employer shall furnish to each of his employees employment and a place of employment which are free from recognized hazards that are causing or are likely to cause death or serious physical harm to his employees.

6. When BBS is in place, focus moves away from comprehensive traditional safety and health programs (despite behavior-based safety company rhetoric).

7. Observers are not adequately trained to identify unsafe work conditions.

8. Behavior lists are compiled with inaccurate accident reports and are too generalized. People developing checklist are not trained to identify root causes.

9. Goals of number of observations per week takes focus away from other safety issues.

10. Management will allow great union involvement in BBS, but not in other traditional health and safety programs.

11. Our [union] goal is to see that workplaces, jobs and equipment are designed in ways that recognize that possibility [of human error] and assure that dire consequences will not result from inevitable human error. The emphasis on workplace and job design must be the same as the emphasis we seek for ergonomic hazards: fix the job, not the worker!

12. Workers and unions are an afterthought to the process and need to be included much sooner to help determine what type of program would be best for the site.

13. 'Proper body position' has become a replacement for a good ergonomics program and well-designed work stations.

14. Incentive programs that focus solely on Lost Work Days or Reported Injuries.

15. Behavioral psychologists usually have very limited work experience – their approach is academically correct, but fail to translate to real-world organizations.

16. BBS programs are based on data and research done in the 1930s by Heinrich, who based his research on reports written by corporate supervisors.

17. Use current workers to identify hazards rather than employing health and safety professionals.

18. BBS programs don't support the hierarchy of controls. Or they turn the hierarchy upside down and focus on PPE rather than the elimination of risks.

19. BBS sidesteps the safety and health risks associated with increased line speeds, work duties, mandatory overtime and other forms of work restructuring.

20. Management should be prevented from forcing any bargaining unit members from participating in this [BBS] program unless they volunteer.

21. Behaviorists believe consequences are the driving force to changing people's behavior. Punishment decreases the probability an unsafe behavior will be repeated.

22. Behaviorism denies internal processing that goes on in human beings.

23. Using positive reinforcement causes resentment between management and workers when employees come to realize they are being manipulated while the safety system stays the same.

24. The simple idea that unsafe actions by workers are the driving force creating incidents and accidents is now outdated. This technique of focusing on the event (the unsafe action) in the hope it will lead to understanding why it happened is doomed to failure. The reality is, the system creates the behavior of both managers and workers. We know you can replace an entire workforce with new people and they will produce the same results. This is true for quality and safety.

Approaches to Ensure BBS is NOT Used for Discipline

When workers ask about a new BBS process, "Am I going to get in trouble or disciplined for the data on these observation forms?" The answer should be, "Absolutely NOT!"

If BBS is used for discipline, it hurts or kills the process. If a worker is asked to observe his friends at work and he discovers that every time he marks a "Concern" his friends get in trouble, what do you think he is going to do? He'll either quit being an observer or quit marking concerns.

In Lean Behavior-Based Safety, we recommend the following four layers of protection be built into the process to make sure that doesn't happen:

Level 1: Management Commitment - We explain to managers in their training that using BBS for discipline will hurt or kill the process, and they commit to make sure that does not happen. They have invested heavily in this process and want to see it succeed.

Level 2: No Names - This is an anonymous process. We do not write the names of those observed on the checklist so no blame can be placed. We don't care WHO. This process is all about WHAT and WHY: WHAT risks are being taken and WHY they are being taken. If any worker is being influenced to take a risk, any other worker doing that job will be impacted by the same influences.

Level 3: Checklist Items are Not Rules or Procedures - This way, even if you don't do the things on the checklist, you haven't broken a rule or failed to follow a procedure. The checklist precautions are VOLUNTARY.

Level 4: The Data Does Not Go to Managers or Supervisors - Instead, it goes to a team of workers who make up the BBS steering team. The data can then be utilized for safety improvement. If management help is needed to make the improvements, the steering team can share the supporting data to substantiate the need without compromising the anonymity of the data.

Chapter Questions:

- If you are planning to implement BBS, do your leaders have any of the common misconceptions about what BBS is or how it works?
- If you have an organized labor force, are you aware of union objections to BBS and are you working to avoid creating resistance?
- Can you communicate to the workforce how you are taking four steps to ensure BBS will not be used for discipline?
- If you have an existing BBS process, can you review these lists of issues with your steering team or committee to find ways to further improve your process?

Part 2: Behavior-Based Safety – An Evolution Was Inevitable

Chapter 3
The Evolution of Lean Behavior-Based Safety

After many successful client engagements and discovering more effective approaches to BBS, Terry Mathis introduced the concept of Lean Behavior-Based Safety to the world.[2]

There, he wrote: The business climate has drastically changed since 1984. Behavior-based safety (BBS), in general, has not! Even a proven technology with documented results such as behavior-based safety must eventually evolve with the prevailing business climate. The traditional behavior-based safety process is fat and out of touch with the realities of today's worksites. Sites considering behavior-based safety are concerned about both the internal and external costs. Sites that have already implemented are straining to provide the resources necessary to continue the process. Other sites have decided not to implement because of the costs and inefficiencies. A leaner approach that remains true to the original principles has proven to be the answer to all these problems.

Problems with Traditional Behavior-Based Safety

Traditional behavior-based safety grew up in a time when many companies still had a full staff. Early BBS processes involved as many people as possible in an attempt to maximize employee ownership and participation. Many of the founders of behavior-based safety utilized resource-intensive techniques such as overtraining, inside-out cultural change and high levels of employee involvement to boost their probability of success. The whole thing worked. It was effective; but it was not efficient.

Behavior-based safety had another problem that did not manifest itself immediately; it was amateur. In the zeal to "empower" employees, behavior-based safety entrusted every aspect of the process to workers who had only minimal training to do technically demanding tasks such as leadership, identifying behavioral targets, coaching, behavioral observation and data analysis. Teams of workers did remarkably well given the challenge, but many opportunities for further gains were missed. The heavy reliance on employee involvement was done purposefully to get the maximum impact on the site culture, but it resulted in other problems. Behavioral targets were not expertly identified. Feedback was not given effectively. Observation strategies ignored good sampling technology. Observation data often contained rich leading indicators of upcoming accidents and their underlying causes, but the data was not expertly analyzed and utilized. Problems remained unidentified, identified problems were not shared with the proper problem solvers and organizations missed countless opportunities to learn how to prevent future accidents.

Most managers were kept distanced from the process in the name of employee ownership, so most organizations didn't even know what they were missing. The results were slow, but improvement was noticeable. We all lived with the inefficiencies because accident rates were decreasing, behavior-based safety was new and we had not yet discovered alternatives.

These problems were not universal. Some sites developed the expertise to run their processes and analyze their data well. Some sites utilized their experts as facilitators or resources to their employee committees and others simply had gifted workers. But an alarming number of processes failed or plateaued due to lack of internal expertise.

Changes in the Business Climate

Since the early 1980s, the business climate has changed significantly. Most sites have experienced dramatic downsizing and re-engineering, and are beginning to adopt new practices, such as lean leadership and lean manufacturing. The manpower available for anything other than production in industrial America is at an all-time low.

During this same period, labor unions saw some of the more poorly-implemented behavior-based safety processes and decided management was using BBS to abdicate its safety responsibilities and simply blame workers. They also noted isolated cases of discipline and punishment attached to behavior-based safety observations and decided it was wrong to ask union members to "spy" and "snitch" on other members.

Today's business climate is far from an ideal environment in which to practice traditional behavior-based safety. The startup time is too long, the external costs are too great, unions resist the process and the internal resources needed to maintain it are simply not available in many companies. This leaves us with three choices:

1. We implement behavior-based safety because it's the "right thing to do" and eat the costs.

2. We abandon BBS and label it as desirable but too costly.

3. We use the fundamentals of the behavior-based safety process to build a lean model to fit today's realities.

Opportunities for Making Behavior-Based Safety More Lean

If we examine the body of behavior-based safety, we find several spots where the "fat" is evident:

- Training – Most behavior-based safety processes require many employees take many days of training to learn and start the process. The strategy of overtraining has to go. All training has to be delivered in an efficient manner, minimized, with only enough philosophy to support the basic principles and a lot of "step 1, step 2" mentality. Training must be focused and shortened for maximum effect in minimum classroom time. It must be memorable, delivered just-in-time and reinforced through non-classroom techniques.

- Leadership – Most behavior-based safety processes are led by teams of employees. This team or committee is often the target of the overtraining, wasting countless amounts of manpower. The team is sometimes used for design purposes to help make the process more site-specific. The team is asked to interpret the data from the observations and recruit and train new observers. All of these tasks require expertise many teams lack. Teams can be replaced with facilitators or smaller teams, which can both decrease the number of people in training and the overall training time, and increase the expertise of the smaller group or individual. Using site personnel who are already expert in some or all of these tasks can also lead to greater integration of the behavior-based safety process into the site structure and management culture.

- Subject-Matter Experts – The focus should not simply be on using fewer people, but on using the right people with the right skills. For example, most sites have someone with data-analysis expertise. Why not utilize this person to analyze data or to facilitate the team?

- Observations – Most behavior-based safety processes recruit between 10 and 100% of the workforce to perform observations. Gathering data is combined with giving feedback in every instance. The number of observers can be drastically reduced and feedback can be focused only in areas where it can make a difference. The observers can do S.W.E.E.P. (Seeing Without Explaining to Every Person) observations that give all the advantages of traditional "upstream" metrics without the outrageous expenditures of manpower. The fewer observers can be better trained and many workers who have hesitations about conversing with their fellow workers about safety issues can be spared the pain. The few people with good coaching skills can be used for focused feedback. The whole process becomes both more lean and more expert.

- Focus – Checklists in many traditional behavior-based safety processes possess 20 or more "critical" behaviors. Observing and giving feedback can become very time-intensive. And long checklists can create a dependence on the observations to maintain the consistency of behaviors. When the frequency of observations goes down, workers tend to quit taking checklist precautions. Shorter checklists require less time to observe and gather data, create habitual competence, minimize dependence on ongoing observations, are more easily remembered by workers, and tend to produce quicker and more focused results. They also take a lot less manpower.

- Data Distribution – Much of the data generated in traditional behavior-based safety is seen only by the steering committee or leadership team. The data could be better analyzed at the management level, or even outsourced. Many world-class safety organizations have reduced accidents to very low-probability risks that often repeat at intervals marked in years rather than days or months. These accident cycles and repetitions are only recognizable in large sets of data. Often, this is best done at the corporate or even multi-corporate level. The data managed by employee teams rarely sees this kind of analysis, and many lessons that could prevent disastrous accidents are never learned by corporations.

Other Opportunities: The Internal-Consultant Approach

Another "lean" technique is to implement behavior-based safety internally without relying completely on outside consultants. The availability of do-it-yourself (DIY) materials for behavior-based Safety has been lacking. Real training and resources for DIY BBS is a new technology that is badly needed and whose time has come.

Lean Behavior-Based Safety is a good alternative for sites with union resistance to traditional behavioral safety programs. The lean process eliminates management omission and can minimize or even eliminate using union members as observers. SWEEP observations can be done by safety professionals or representatives.

Sites that have already implemented behavior-based safety can use lean techniques to put their own processes on a diet. Checklists can be focused to fewer behaviors. Leadership teams/committees can begin to downsize through attrition or in a more accelerated manner. Observer teams can be supplemented with SWEEP observers and eventually replaced. The best traditional observers can become safety coaches sent to the "hot spots" identified by SWEEP observations. Data can be redistributed or even outsourced for analysis and distribution in the organization. Many sites have found that the diet not only helped their behavior-based safety process to reduce the use of resources, but actually re-energized the process. New, leaner processes are being implemented or retrofitted in many US firms, and the trend is spreading to other parts of the globe.

Case Study

Sites have been implementing this leaner version of behavior-based safety since 2001 and have, in general, gotten equal or better accident reductions than sites implementing traditional behavior-based safety. But there wasn't a study of side-by-side implementations at the same site until 2004. Two manufacturing sites implemented two simultaneous behavior-based safety processes, one traditional and one lean. All the initial training and design was completed by year's end in 2003 and the observation process began shortly after the first of the year in 2004. The sites using the lean approach achieved slightly better results with significantly less use of both internal and external resources. The first-year statistics are in table 1.0 below.

Item	Site 1 Lean	Site 1 Traditional	Site 2 Lean	Site 2 Traditional
Population of Test Division	321	292	222	239
Leadership Team Membership	4	12	4	12
Number of Observers	11	29	9	24
Employee Hours for Implementation	220	1181	188	1176
Days of Outside Consultants Time	7	19	7	19
Hours/Month to Run the Process	30	164	22	144
Beginning TRIR (3-year average)	2.8	3.1	2.4	2.9
Year One Reduction in TRIR	51%	41%	49%	43%

Table 1.0

Conclusions

Those that have opted out of the behavior-based safety trend because of expense or resource requirements now have new options. Firms that have traditional BBS processes already in place have a way to reduce manpower requirements without sacrificing effectiveness. The leaner version may be a better fit for small sites, sites with limited budgets, and/or sites with inadequate resource availability. Simply using parts of the technology without opting for the whole process may prove effective for those with specialized needs, difficult logistics and cultural complications, including union resistance. This new way of thinking about behavior-based safety has brought a useful technology into the realities of today's business climate.

What Can Safety Learn from Lean?

If you think Lean is only for manufacturing, look it up on Wikipedia. You will find that Lean principles, Lean thinking and Lean tools have been adapted and applied to everything from service industries to software development. In a highly-competitive global marketplace, almost everyone is looking for competitive advantages. Lean is not just about less; it is about efficiency. It does not seek to do as well with fewer resources; it seeks to produce excellence by focusing resources on highly effective activities while eliminating the activities that do not add value. The basic premises of Lean offer some potential opportunities to further improve safety as well.

Most people associate the term Lean with the Toyota Production System. This combined management and production system helped a small company grow to world-class size and market share. As they did so, many auto manufacturers and other industries studied their methods and tools to learn to improve their own organizations. The system is multi-faceted, but several tenets within it have good potential application to safety: customer orientation, focus on value, efficiency through elimination of wastes, questioning existing wisdom and continuous improvement.

Customer Orientation - Many safety professionals view their customer as upper management, the board of directors and/or stockholders. Others don't think of their safety process as having a customer at all; it is simply an aspect of management and a service to the organization. Lean thinking would point to the worker as the customer of the safety process. As such, the process should seek to better understand and meet the needs of the worker rather than seek to install safety programs through command and control. Many organizations view the worker as the "problem" rather than the customer. The goal of such processes is to limit the ability of the worker to take risks and therefore reduce accidental injuries. Very few safety programs market themselves to the worker and seek to "sell" them on the process' merits and worth. It is assumed that safety is a duty of an employee and compliance, rather than excellence, is the goal. These processes tend to evolve negatively-oriented goals. Safety becomes an elimination of accidents rather than a strategy for excellence. The goal is not so much to succeed as to not fail.

Focus on Value - In Lean Manufacturing, value is defined as any action or process that a customer is willing to pay for. Obviously, people choose to pay for products or services they like or need. They base their selection on availability of features and aspects they prefer over the alternatives. Organizations traditionally spend a lot of resources on market research to find out what their customers value and to make sure their products and services have those features. In safety, we tend to focus on what the organization wants, and ignore the wants and needs of the customer. Certainly, if you view the worker as the problem rather than the customer, this makes sense. However, viewing the worker as the customer opens a whole new avenue to determining the most effective way to design safety products and services. How many workers are willing to pay to attend safety meetings or training sessions put on by their employer? How many would choose another type of PPE or tools or equipment to do their jobs if their input was sought before making those choices?

Efficiency Through Elimination of Waste - In manufacturing facilities, many practices and designs become customary and remain in use long after the reason for them changes. As Toyota began to examine their processes and practices, they found steps and designs that no longer added value. The most prominent of these was called Muda (an activity that is wasteful and doesn't add value or is unproductive). Unfortunately, many of these antiquated practices still used worker time and effort while adding little or no value to the process. Toyota began to eliminate non-essential transportation of products, inventory that was not immediately needed, process steps that duplicated tasks, and similar activities. The two less known categories of wastes were Mura (unevenness or anything that interferes with even flow of processes) and Muri (overburdens such as too many or overly difficult activities). Toyota sought to redesign processes to maximize flow and even out the burden placed on workers by the design of their jobs.

In safety, we often have traditional activities that take up time but fail to add value. We have our workers attend repetitive and boring training that keeps us legally compliant, but does not make us safe. Our accident investigations tend to fix the blame, but not fix the problem. Traditional approaches to behavior-based safety include massive overtraining, resource-intensive overuse of observations, data analysis by employee teams with no statistical training and checklists with so many behaviors they overburden workers rather than empower them to progressively change a few habits at a time. Safety elements of new-employee orientations are also overloaded, more designed to avoid liability than to avoid accidents. Many safety programs "go through the motions" without asking if the motions add value.

Questioning Existing Wisdom - Toyota began to question the assumptions of Frederick W. Taylor and Henry Ford and practices such as time-and-motion studies. This was almost a sacrilege in the day when these men and their techniques had created and defined the industry. However, as workers became more educated and capable, Taylor's ideas of breaking jobs into tasks and sub-tasks became demotivating and routine. Ford's ideas of mass production also evolved, and newer and better methods began to emerge. Quality issues caused Toyota and others to question their processes and find the cause of defects that produced scrap and other wastes. Rather than simply look for Kaizen (ways to make small, continuous improvement), Toyota began to also question whole methods and processes, and look for Kaikaku (large, transformational improvements).

Like the automotive industry, safety has its founding fathers and revered pioneers. Safety programs have been fashioned and duplicated based on their theories. New approaches to safety have been built on a blind acceptance of their assumptions. Recently, however, we have seen many of these "sacred" theories and practices questioned, and new thinking has sprung out of their ashes. Safety management practices have evolved from command-and-control to worker engagement. Safety focus has evolved from conditions, to behaviors, to influences on behavior, and from compliance to culture. The assumptions of Heinrich and others have come under new scrutiny and the common practices such as behavior-based safety and safety rewards and incentives have been questioned. A much more practical and less academic attitude toward the traditional wisdom of safety practice is taking shape. Safety practices and theories can't just "make sense", they have to prove themselves and demonstrate they can work in the real world.

Continuous Improvement - Just after World War II, while much of the world was perfecting phrases like, "If it isn't broken, don't fix it," Toyota was promoting the idea there is no dichotomy such as broken versus Fixed, and that anything, no matter how good, can always be better. They challenged their shop floor workers, not just the managers, to contribute ideas for continuously improving every product and process. They adopted W. Edward Deming's advice that the people closest to the work often know the most about it and problem-solving is best done at the level with the most expertise. While the average factory production worker in other parts of the world made one or two improvement suggestions per year, Toyota workers were making close to one hundred. Many of these ideas were adopted with great results. Workers were cross-trained to do any job in their area and the improvement suggestions increased. Workers became the primary source of improvement ideas, and the quality and productivity of the factories improved continuously.

In safety, we are also evolving away from having managers and safety professionals make all the decisions and improvements to asking workers for ideas. Many safety suggestion systems are poorly designed and cannot handle the volume of input they receive. We are not yet highly effective and efficient at managing worker safety suggestions, but we are beginning to see the value of doing so. More and more organizations have begun to ask for worker input and are perfecting the processes of handling this wealth of ideas. The whole buzz about safety culture is an indicator that we're turning away from the old idea that safety must be "managed and supervised", and realizing worker ideas and safety culture are the keys to continuously improve safety. Organizations are going beyond asking for input and are implementing worker-driven and worker-led processes to improve safety. Safety committees which were traditionally all managers are beginning to have hourly workers as regular members. Organizations are realizing managers can take safety performance from bad to good, but the workers and the culture must be engaged to make it excellent.

For many, Lean is not so much a set of concepts as a set of tools to identify and eliminate waste or inefficiency. The tools are often better known and utilized than the concepts. Even organizations out of the mainstream of Lean often utilize Value Stream Mapping, 5S, Kanban walks and versions of poka-yoke (a system for eliminating and preventing errors). While the Lean tools can be helpful, it is Lean thinking that has the greatest potential to begin a significant change in safety performance. All meaningful change begins with thinking differently. As Lean concepts become better known and understood, perhaps they can be better utilized to improve safety at the strategic level.

Once you get past the stereotype of Lean Manufacturing and consider the concepts of Lean from a more generic point of view, they can be applied very well to safety. Accidental injuries are defects in our processes. We can improve the quality of these processes, thus diminishing the defects, by making them more efficient and removing the waste. This will demand we question and examine the existing process norms and the theories that drive them. We must decide who the customer is and what the customer values. We must design safety like a great product that gives the customer both what is wanted and what is needed. Safety and Lean can forge an alliance to reduce the greatest of all possible wastes: accidental workplace injuries.

Lean does not mean less. Rather, it is a focus on value-add and the continued identification and removal of activities and efforts that add waste; or more simply put: greater value through less work. As a result of the popularity of Lean Methodologies and Lean thinking, a natural evolution to BBS was not far behind.

Opportunities for Lean

Integration - One of the most effective ways to minimize resistance to change is to minimize the perception of it. What existing structure, process, team or cultural capability would produce a shorter path to results and create the impression of evolution rather than revolution? Perhaps the 5S team becomes the 6S, or integrate your focus on quality to include safety with behavior-based quality & safety? Both have proven successful in the right environment. What about an existing team, sub-team or committee?

Another form of process integration is to combine Lean BBS with other compatible processes. Since many other processes, including manufacturing, already utilize Lean, the marriage of such processes with Lean BBS is a good fit. One pharmaceutical company innovated a new process utilizing Lean BBS for both safety and quality. Because their product quality is a matter of public safety since a contaminated product could harm its users.

Case Study in Lean BBS

In 2003, AstraZeneca Westborough (AZWB) planned, developed and implemented the BEδT* (Building Excellence in Safety & Quality Together) Program. BEδT is a far-reaching, behavior-based program aimed to improve two vital aspects of a pharmaceutical manufacturing site: SHE and quality compliance. The foundation of AZWB's BEδT program is an innovative partnership between two unlikely organizations: Quality Assurance (QA) and Safety, Health and Environment (SHE). SHE and Good Manufacturing Practices (GMP) are core to AstraZeneca's business, and we believe BEδT will ensure our employees never lose sight of our aspirational goal of zero accidents and incidents (Target Zero) and Right-First-Time (100% quality compliance). The program has required a team approach and active participation by employees from diverse backgrounds.

There are several reasons why AZWB chose to invest in the BEδT program. Some of these reasons will become more apparent after first taking a step back to review where we've been as a site and the challenges ahead.

The Greek letter "δ" (delta), used commonly to represent change, was chosen as a hybrid of the combined "S" & "Q" in our acronym. Ironically, Westborough's BEδT program was defined and named prior to AstraZeneca's "Being the Best" campaign.

Since the Astra - Zeneca merger in 1999, the AstraZeneca Westborough supply site has undergone a tremendous transformation, not only in our overall business performance, but in our SHE performance as well. Statistically speaking, AZ Westborough has demonstrated exceptional SHE performance improvement over the past three or four years. While this is great news for our employees and the site, it has presented the SHE team with an interesting challenge. How do we maintain or improve on the current performance trend? And from a practical standpoint, do the statistics reflect reality? Finally, on both levels, how do we demonstrate continued SHE performance improvement going forward? What, if anything, do we need to do differently?

When further defining this issue, the SHE team in Westborough recognized:

- We were at a point where statistical SHE performance improvement was starting to plateau,
- A minimal uptick in injury or illness incidence at first glance may have given the appearance of declining SHE performance,
- From a practical standpoint, we were confident there was still a good deal of room for improvement,
- While SHE remained a key element of our culture at the site, it was not yet fully integrated into the business,
- Initiating any new SHE programs as stand-alone initiatives in the midst of good SHE performance may have not been well received by employees or may have been perceived as "just another SHE program",
- And more importantly, we needed an innovative approach.

Reaching our vision of being the best in SHE meant taking a new approach - recognizing the possibility that incremental improvements in performance could not be achieved using the same methods and thinking that gave us the level of performance we have today. The way forward involves a more innovative, business-integrated, customer-focused mindset. Success will only be achieved when SHE is perceived by our internal and external customers as a true business partner. The way forward will be likened to an overtime period in a ball game, when winning is derived from renewed energy, exceptional teamwork and even a new strategy. Anything less gets you no better than a tie game.

With all of this in mind, in 2003, we took some important steps to position ourselves for achieving our vision for being the best in SHE. To be the best in safety, we've implemented the BEδT program, a combined safety & quality compliance behavior-based process. To be the best in health, we've taken a collaborative approach to both educate employees on AstraZeneca products and managing health issues. To be the best in environmental performance, we've mitigated the risk of pharmaceuticals in the environment while saving energy. In the following, we will focus on our behavioral approach in safety and quality – a new way of operating which has put us on the path to being the best.

The traditional approach to improving safety has been to conduct an investigation (after the fact) to determine the cause of an accident. To illustrate this, consider an analogy of an employee falling from a cliff. In this analogy, an accident investigator would pose such questions as, "Why did the employee fall off the cliff?' and "How can we prevent that from happening again?" Our measure of success going forward was defined by the number of employees who fell off the cliff, a lagging performance indicator. To be the best in safety, we need to look beyond this traditional approach. BEδT is an effort to do just that.

How do we change the probability of someone falling over the edge? We need to gain a profound knowledge of our behaviors. We need to move to the top of the cliff and measure how many employees get too close to the edge and, more importantly, determine why they were too close. In doing this, we are measuring at-risk behaviors before they become accidents.

The same traditional approach exists in performing quality compliance investigations, i.e. conducting an investigation (after the fact) to determine the cause of a process deviation or problem. This presented an opportunity to partner with QA to use a common method for achieving our mutual goals – target zero and right-first-time.

To assist with the design and implementation of a customized solution, AstraZeneca solicited proposals from a handful of the top behavioral safety specialists in the US. One of our key goals was the willingness of the organization we chose to take a truly 'custom' approach. We felt strongly that an off-the-shelf process would not work well at our site. After several weeks of review and discussion, AstraZeneca chose ProAct Safety, Inc. ProAct Safety offered the right balance of flexibility in approach and value.

When AstraZeneca first contacted us, a strategic committee had already considered the possibility of combining quality and safety into one initiative. It was obvious this was not just another site looking to reduce their accident rates. The site leaders had clear goals, solid expectations and were seeking a partner to test the feasibility of their vision and turn it into a successful reality.

We quickly determined this project would not only stretch the boundaries of traditional behavior-based safety, but would challenge the very model of consulting under which such processes are typically delivered. The challenge was more than offset by opportunities for innovation and the even more elusive opportunity to form a true collaborative partnership between a client and consultant that exists more in marketing materials than in the real world.

The flow of implementation was a modified BBS process flow, but the steps had significant variances. The make-up of the steering team straddled safety and quality employees.

The checklist development involved review of massive amounts of quality-assurance data to determine the behaviors that most impacted right-first-time goals and outcomes. The steering team spent time considering whether or not it was feasible to observe for both safety and quality data at the same time in all locations. For this and other reasons, the steering team elected to perform sample observations for two months before finalizing their observation strategy.

Even then, the newly selected and trained observers were required to bring steering team members with them for the first month to ensure the strategy was followed and working. When the observers were rotated, steering team members accompanied the old and new observers for another month to help smooth the transition and maintain the integrity of the observation strategy.

During the first year after process kickoff, the site manager was reassigned and the site received a new general manager. He was not only an acceptable replacement, but even expanded support among managers with new insights on how the process could be more successful.

The first step in the process was to perform an organizational assessment to determine the site's readiness for using a behavior-based approach. Focus groups, perception surveys and existing data analysis revealed AZWB had an ideal climate for accepting behavior-based solutions.

Two teams were formed to design, develop & implement a customized process for the site. A strategy team set objectives, defined critical success factors, allocated resources and established boundaries for developing the process. A steering team, made up of members who met specific leadership criteria, was chartered to customize and implement the plan. Process advisors from Quality and Safety provided ongoing coaching & guidance to this steering team. Consultant training centered on this later team who learned the process and designed tactical elements of the process.

Early communication and frequent briefings were conducted to ensure managers and supervisors understood the overall scope and resource needs of the process. As managers learned the details of what their steering team had designed, they engaged in defining their own support behaviors to ensure the success of the process. Management buy-in was critical to the success of the process.

The steering team training and development curriculum included basic team skills, introduction to behavioral science, process design, data analysis and problem solving, observer training and navigating long-term process issues. Over 80 hours were devoted to ensuring the team had the competence and confidence to effectively implement and manage this process.

A "BEδT Practices Checklist" identifying the site's critical behaviors was developed by analyzing two years of accident & quality data. The team piloted the process so we could better understand the observation-feedback element. Their experience enhanced their ability to train and mentor the initial observers, known locally as "BEδT Buddies."

Site-wide employee briefings were provided to coincide with the start of the pilot. The key objectives of these 2-hour sessions were to:

- Introduce BEδT and the behavior-based methodology;
- Present the business case for implementing it at AZWB;
- Help reinforce that observations are anonymous – they are only looking for the "what", "how", and "why", not the "who";
- Demonstrate the importance of feedback in understanding & changing behaviors;
- Solicit volunteers to be the first group of BEδT Buddies.

Thirty-seven BEδT Buddies were screened and selected, and represented all shifts and major areas at this site. They initially received 4 hours of formal classroom training, which is typically the base training for observers. However, based upon the challenges encountered during the pilot, the steering team implemented an innovative solution by providing intensive on-the-job training and mentorship for the BEδT Buddies.

During the first 24 months over 7,500 observations were logged involving over 21,000 people observed that provide a solid baseline from which to focus continuous improvement efforts. Of the seven categories contained on the checklist, four behaviors (2 quality and 2 safety) were identified early on as "quick wins" – defined as having a measurable impact and being achievable within a short timeframe. The team communicated these initial findings to the site and developed action plans to address them.

After the first 12-18 months of working the process and to ensure continuous improvement, there was a need to refocus the efforts. In early 2005, the team undertook an effort to narrow the focus of the overall process. The critical behaviors checklist, which served well in the early days of data collection, was in need of updating.

Safety and quality data from the most recent year and a half was reanalyzed and used to shorten the critical behaviors checklist. The site was able to eliminate some behaviors in areas where they had sufficient data and were observing consistently high percentage of best practice, in lieu of some focused observations. For safety improvement, critical behaviors associated with lacerations and glass handling was added and greater focus was placed on aseptic technique, for quality compliance.

During the past year, we have begun to use the observation process for discrete projects with some great success. During the site's annual maintenance shutdown, a four-week time period where all production and systems were taken down, refurbished and updated, the process was used for contractor safety surveillance.

A special critical behaviors checklist was developed, mirroring OSHA's top construction safety issues. AstraZeneca project managers took on the role of observers using this checklist, and were able to provide weekly snapshots of individual contractor team performance opposite these behaviors. At first, the use of this process was met with some skepticism. It was new to most of the contractors. It quickly became somewhat of a contest among them to see who could end up with the highest % of best practices in the weekly lunch meetings held with the superintendents/foremen.

Also, we are using the process for large capital construction projects (new microbiology lab). This approach was certainly new to our construction contractors, but is now well received and a useful, real-time method for identifying concerns and putting actions in place to prevent incidents, mishaps and issues which can be costly and time-consuming."

The process had a great initial impact on the site's performance, reducing accidents with injury by 50% and contributing to a 17-point rise in pristine batches, a measure of quality right-first-time. In recognition of these achievements, AstraZeneca awarded the Westborough site the Chief Executive's SH&E Award.

Case Study: Client Experience - 5S to 6S? No Way!

Having combined several Lean BBS processes with other efforts, we always look for opportunities to do so in each new implementation. We were working with a client on the West coast whose workers were hesitant to take on another new process. They had been very successful in their recent 5S process and took great pride in what they had accomplished. We suggested one alternative to introducing new infrastructure and processes is to bolt on Lean BBS to 5S. We could call it 6S and utilize similar methodology to address crucial safety behaviors.

The site steering team was adamantly against the idea. They felt it would dilute their 5S program rather than build on it. They also noted other workers had participated in steering 5S and would feel like this new team was stealing their process. At this point, we decided not to change the 5S program but to take another approach which proved successful and did not create resistance. What works at one site does not necessarily work at another. Being flexible and listening to those in tune with the site culture rather than just following past practices or first impressions is crucial to success.

Other Opportunities to Utilize Lean Thinking and Methods

In addition to the ideas in this section and the case studies above, there are several other areas in which BBS can benefit from Lean, including:

Training - Overtraining is a common problem in many BBS implementations, especially when the focus is on implementing a methodology rather than recognizing rapid and sustainable results. Consider: what information is needed at this time to accomplish results in the next 90 days? How can you build on that once success is recognized? Why teach a team how to analyze data and develop action plans before they have their own data? Think just enough and just in time.

Steering Team Size - A team that can never all get together isn't a team. A successful process needs a team representing the interests of the population to make important decisions of improving process and safety. What that team looks like, how many are on it and the makeup of levels involved is culturally-specific and based largely on trust.

Checklist size - Effective BBS processes determine the most important behaviors to focus on based on historical incidents and observed common practice. A focus on more than five or six items is not a focus. Moreover, centering attention on the wrong things is not very efficient. If your checklist isn't focused, neither is your process. The checklist needs to adapt to fit the needs of the process. A process should never revolve around a checklist or be dependent on others to print and modify it.

Observation Strategies - Blanketing observations is a tactic to capture a sampling of all common practice. Too often, this becomes the only observation plan. There are four other common observation methods: Aimed, Blitz, Seeing Without Explaining to Every Person (S.W.E.E.P.) and Self. Depending on the data and value that needs to be derived from the process, the observation strategy should evolve to focus on new results. Is it aiming at the right thing, time of day, etc.? Is it blitzing where it is most needed based on identified influence? S.W.E.E.P. observations are often used to validate a high percent-safe finding, and self-observations are often used with logistically challenged workforces.

Data Utilization - Far too many processes do not analyze their data on a frequent enough basis to understand how to prioritize action plans, address data quantity and quality, identify trends and influences on behaviors, and respond to these influences. Any mature process combined with a healthy safety management system can indicate where, when and why the next incident will occur with a high degree of accuracy, based on observed practices.

Voice of The Customer - When a process is launched, all those impacted should be briefed to begin addressing the "What's In It For Me" (WIIFM) question. Realize, however, the question never goes away; the answer will change over time. This process must treat workers like the customers they are. If process leaders are not communicating in a way that creates strong knowledge and value recognition (e.g., everyone can name the behaviors of focus and recent successes), the support will wane.

Documented Roles, Responsibilities and Results - For decades, change initiatives have experienced the middle-management backlash, due to bottom-up and top-down approaches. BBS has been no different. We desperately need supervisors to support the process, yet most processes fail to explain and develop approaches to ensure this occurs. The roles, behaviorally-defined responsibilities and results expected need to be developed for all involved, from those observing to those supporting leading (e.g., integrate with leader standard work) and those serving only in a customer capacity.

Ultimately, the goal of BBS is to improve safety, help those impacted by it to make safer decisions and address the factors that might influence risk-taking (e.g., perceptions, habits, obstacles and barriers). While there is certainly much more to the Lean BBS methodology, consider how you can use this Lean BBS thinking to eliminate unnecessary or non-value added steps and maintain a results-orientation rather than a crank-the-process one. Your focus and attention should be spending your time to improve your process and using it to improve safety.

Summary: What it Takes to Make Lean BBS Work

- Readiness
- Customized Process
- Leadership Support
- Supervisor Cooperation

- Process Leadership
- Trained Observers
- Worker Support
- Data Management Strategy
- Succession Planning
- Proactive/Positive Accountability Plan

Chapter Questions:

- Who is the best person to determine site readiness?
- Do you have an existing safety committee or team that could be used to steer the BBS process or will you need to form one?
- Do your leaders have enough knowledge about BBS to support it?
- Who should train observers, and should they be mentored or coached on their first few observations or sent out in pairs?
- What is the best way to schedule worker briefings to ensure you get everyone to attend with a minimum number of sessions?
- Can you purchase software to manage your observation data or will you be required to develop the software with your in-house IT department?

Part 3: Owning the Ability to Implement and Improve

Chapter 4
Who Leads Effort?

Through having personally implementing hundreds of new processes and improving just as many efforts begun by other experts, we have learned that BBS may need to be structured differently for every company.

Socrates advised, “Prescription before diagnosis is malpractice.” Like any change designed to positively improve a culture, the approach should be made to fit the operational and logistical realities, and should consider what people will and won’t support. One area of potential malpractice in implementing BBS concerns defining who is responsible for overseeing the efforts, from design through monitoring and continuous improvement. Some say the efforts should be led by someone with a full-time responsibility, while others insist this should be shared by a group such as a steering team or committee. Both approaches have merit and should be explored further.

Facilitator, Sponsor, SPA?

If a company is looking to implement across many different locations and cultures, often a facilitator or internal consultant is used to lead the efforts. This person must have a firm understanding of the options and what leads to success or failure. In some very small companies or groups, the full responsibility of BBS may be assigned to a single person because of resource or logistics limitations. Most often, the role of facilitator is not a full-time responsibility. Typically this individual takes on the responsibility in addition to their regular duties, slowly enables a team of individuals to become self-directed and, over time, disengages from having a hands-on role to allow ownership to exist within the group.

The facilitator will need to have a consulting mindset to determine the best approach, and have a teachable point of view to know what methodologies to choose. The facilitator will also need to explain these methodologies to those impacted by the process in a just-in-time manner. Sometimes this person is someone from the safety department, but not if there would be a conflict in current area of focus (i.e., currently focusing on compliance or enforcement), or how the safety professional is currently viewed (think safety police versus safety coach). Other times, this is a professional development opportunity for an individual, or a job for someone who has experience leading complex change.

Rather than assigning someone to make all the decisions or lead the bulk of activities, sometimes all that's needed is a group sponsor who makes the decisions. The sponsor would be someone from leadership who can speak to other leaders to gain support. This person would also speak with the employee population or bargaining unit representation to gain support and facilitate involvement. This person can act in a role to help the process get up and running, or even act as a member of a team with shared responsibilities to make the process successful while also ensuring continuity with other traditional safety efforts and the overall operational or business strategy. Some companies identify a Single Point of Accountability (SPA) to ensure someone is directly accountable for leading the efforts, coaching for continuous improvement and removing the barriers to success. This allows everyone to know who specifically to approach for help with safety concerns.

Structuring a Team

The most common approach is to have a team become the designers, communicators and overseers of the efforts. Where possible, it is preferred to attempt to leverage existing teams as long as they don't already have too much on their plate. This team must be viewed as successful with past efforts, be representative of the work areas, levels and culture, be able to meet frequently enough, develop consensus on decisions, and be made up of known influent-

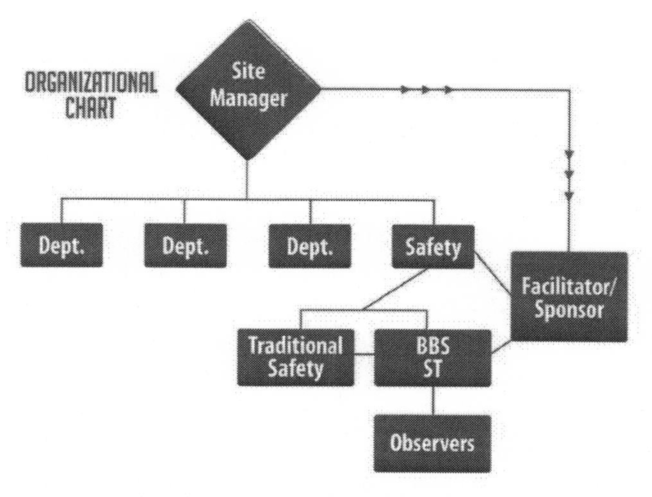

Figure: Sample Common Organizational Structure for BBS Steering Team (ST)

ial people. Culture change happens best from within, so teams should be structured with those who can successfully lead and influence this change.

Sometimes a team already has the responsibility of leading other safety efforts driven by the safety department that are more traditional in nature (e.g., compliance, enforcement, policy and procedure creation, inspections, auditions). Where this is the case, consider developing a new team that will complement the existing safety efforts, rather than taking on everything at once. Resource realities and trust levels typically determine how to structure this and who to involve.

However you lead the efforts within your company, remember these principles. Culture change happens best from within. People are more likely to have ownership in something they had a hand in creating. Self-directed efforts are ideal but don't happen on their own; sometimes a facilitator is needed. There should be continuity between BBS and other safety efforts, and between the overall safety and business strategies. External experts should advise but not direct the approach. The approach should be made to fit the site. If you are unable to modify the structure, terminology, focus or materials, it will eventually become something of value you once did. Consider your organization. How can you best structure this in a way to meet your realities and ensure it delivers sustainable value?

- External Consultant
- Internal Consultant
- Hybrid Approach: Leveraged Use of External SME

Expert Utilization

Unlike the traditional model of using consultants to do all the legwork of culture and safety improvement, we recommend what leveraged use of consultants. In a leveraged consulting project, several things can be accomplished with greater efficiency:

- The use of client personnel talent and cultural savvy can be utilized in implementation planning to better customize the process for the organization.

- Consultant expertise can be transferred to the client personnel with opportunities to develop and demonstrate that expertise with the coaching of the consultant.

- The work of the implementation can be accomplished more quickly by utilization of trained client personnel to multiply efforts.

- Greater participation and ownership of the project can benefit personnel who have active, rather than passive, roles in the design and implementation.

- A plan for expert consultant disengagement can be carried out in which the consultant's role shifts from performance of tasks to subject-matter expert with client personnel taking more inclusive roles. This allows for longer-term planning as internal expertise is developed to carry out process goals with minimal consultant assistance.

- The costs of external consulting services can be reduced by only using the expert consultant for necessary activities and subsequent advisory work without negatively impacting the probability of success.

<u>Internal Consultant</u>

Another option is to have an individual within the organization trained as an internal consultant of the Lean BBS process. This person needs the ability to understand their current culture and what people would and would not support. Therein lies the key to making BBS successful for any company in any country — the approach must be made to fit the site rather than the site being made to fit the approach. Certainly an external expert can gain insight into the culture, but individuals inside the culture already have this insight and simply need to learn how to customize and implement Lean BBS to fit the culture.

An external consultant or company can instruct the individual on specific consultative methods to identify many different customized and creative ways to implement BBS and, most importantly, which paths to avoid. SMEs can also help to modify existing BBS processes to maximize effectiveness. This approach is an option for companies who want to maximize their own ability to implement BBS while minimizing outside costs. Companies using this path strategy should have high-quality personnel and internal resources.

This individual can be certified to return to single or multiple sites and consult on implementing customized BBS processes. This approach is not a straightforward train-the-trainer course, nor is it intended to teach individuals to only simply deliver training on behavior-based safety. Every site will have its own unique challenges and cultures. To allow the internal consultant the most opportunities for success, it is extremely important they understand and internalize the strategies for identifying the site-specific variables that have become, or could become, problematic barriers. What works at one site, many times will not work at another.

This training should fully qualify and prepare an internal consultant to create a customized plan to utilize the latest proprietary Lean BBS technologies for spearheading an implementation. Most importantly, it will prepare them to anticipate and handle issues that can challenge the success of BBS efforts while also looking for the opportunities to minimize the perception of change, achieve the quickest success possible and ensure process sustainability.

Chapter Questions:

- Who is the best person to lead your BBS process implementation or improvement?
- Does this person need training and/or assistance?
- Will this person also be the SPA for BBS?
- Have you selected a person whose future with the organization is stable?
- Is this person well-respected by his/her colleagues?
- Is this person at the right level in the organization to have enough authority to facilitate and provide resources to the BBS process?

Chapter 5

Determining Readiness, Fit and Strategic Planning

Develop a Customized Approach

To ensure the highest level of success, the implementation should be customized to each site to fit each unique culture. Every site has its own unique challenges and cultures. What works at one site will not always work at another. Truly understanding the site-specific variables that have lent themselves to success, or have become problematic barriers, will allow you be the most successful in understanding how to improve.

Site Organizational & Cultural Safety Assessment

The following outlines a recommended path strategy for determining both site readiness and the best way to implement the process at a specific site.

The assessment will include 4 major activities:

1. Document Review
2. Site Tour
3. Interviews
4. Strategic Planning Session

Each of these activities can include a number of issues as outlined below. Internal consultants who are already familiar with the site specifics might choose to limit the assessment issues to those they are least familiar with or those that present potential challenges.

1. Document Review

The consultant will need to review accident, incident and near miss data, if available, for the past three years. Also, OSHA logs or accident trend information for the past 3-5 years should be reviewed, and a discussion should take place about policies and results of drug testing, safety training, safety meetings and supervisor safety responsibilities.

The following is a list of items we typically like to examine in the initial stage of an assessment.

1. Examples of Safety Policies & Procedures
2. Blank Work Order Form
3. Blank Incident Analysis Form
4. PSM (if applicable)
5. Example of Management of Change (MOC) Templates
6. Process Safety Incidents
7. Examples of Safety Communications – Posters, Newsletters, Company Videos, Bulletin Board Postings, etc.
8. Example of Safety Communication Flowchart (if applicable)
9. Breakdown of all Safety Committees
    a. Type

b. Number in Each

c. Tenure

d. Employee Type (Manager, Supervisor, Hourly)

10. Safety Meeting Schedules

11. Annotated Organizational Charts

12. Recent Cultural Measurements

13. Recent Perception Measurements

14. Incident Reports for Past 2-3 Years Including Near-Miss Data

    a. Incident Type

    b. Incident Cause

    c. Incident Description

    d. Employee Work Group

    e. Employee Tenure

    f. Tenure on Task

    g. Shift

    h. Day of Week

    i. Time of Day

15. Accident Rates for Past 5 Years (preferably on a trends chart as well)

    a. Total Recordable/Case Incident Rate

b. Lost Time Case Rate

c. Severity Rate

d. Days Away/Restricted or Transfer Rate

16. Financial Data

a. Other Accident-Related Costs

The purpose of this information is to help determine what factors would support improvement and what factors would challenge, and to ultimately facilitate a site-specific strategy to proceed with systematic integrated improvement. The assessment can also discover the few factors that might make targeted effort virtually impossible so the site can better prepare or look for alternatives.

2. Site Tour

Early in the visit, arrange a site tour for the consultant. This should include all major areas of the site, at least some of which should have work in progress. The goal of this tour is to give the consultant an overview of logistics, tasks and basic safety issues involved in the processes of the site. Issues to look for during the tour include:

- Can workers be observed in groups or will they need to be observed individually?
- Can workers break away from their jobs to do observations without major interruptions to operations?
- How far will observers need to travel to observe other workers and how long will that travel take?

- Are there special events to be observed as they occur (batch processes, outages, line breaks, incoming shipments, etc.)?
- Are there any areas difficult to observe (clean rooms, guard towers, tight spaces, etc.)?
- Are there areas where observers might be at risk performing observations (high-traffic areas, special PPE areas, confined spaces, etc.)?
- Are there workers in some areas or on some shifts with no one else available to observe them?
- How many observations will it take to observe each worker in each area?
- Any other conditional or logistic challenges?

3. Interviews

Individual interviews should be arranged for the site manager, H.R. manager, and direct reports to the site manager. Group interviews should be arranged for as many supervisors as possible and for 10-15% of the hourly workforce. If the interviews are all held in one office, they can be scheduled back-to-back on the hour and half hour. Interviews with key managers may take longer, but take care not to schedule long interviews for workers or supervisors, as they can seriously interrupt operations and tax the attention span of people pulled off the job.

The purpose for these interviews is to determine how the employees at each level perceive the current safety efforts and the opportunities for further improvements. These perceptions will fall into three categories:

- Positive perceptions: those that will aid and support BBS
- Negative perceptions: those that will make BBS difficult or impossible
- Neutral perceptions: uncertainty or apathy about the issue or issues

In addition to the interviews, formal perception surveys can be administered during the interview times. These tend to offer more quantitative information about perceptions and can either validate or challenge the assessment of the interviewer.

The following perception categories should be covered:

1. Organizational and management support for safety
2. Communication of safety information
3. Clarity of safety expectations
4. Responsiveness to safety work orders / suggestions (maintenance and facilities management)
5. Status and effectiveness of current safety programs and processes
    a. Safety staff and professionals
    b. Programs, policies, practices
    c. Training
    d. Meetings

   e. New employee orientation

   f. VPP (if applicable)

6. History of past change efforts

7. Employee willingness to participate in safety improvement efforts

8. Skill levels and past training in key areas

   a. Meeting skills

   b. Teambuilding

   c. Data analysis

   d. Problem solving

9. Use of incentives relative to safety

10. The priority of safety in relation to other priorities

11. Trust between levels of the organization in safety efforts

12. Employee interaction and involvement

13. The use of blame, punishment and discipline for safety offenses

14. Perceptions of the preventability of accidents

15. Organizational and individual focus on safety issues

16. Safety compliance levels and issues

17. Common practice versus directives in safety

4. Strategic Planning Session

An exit meeting with key personnel should occur at the end of the assessment to discuss the general findings, factors that support improvement and factors that could challenge efforts. Additionally, this session should address how the current organizational, operational and process safety issues impact safety success at the plant. Discussions should focus on what extent the site is ready to implement a focused improvement initiative and what customization, integration and change management tactics should be utilized. This should consider any of the following which apply:

- Assessment findings
- Potential cultural and operational issues that may impact the success of current and future initiatives
- Guidelines for specific process issues
- Employee relations
- Key Process Indicators (KPIs)
- Training recommendations
- Ongoing motivational plans
- Integration potential with other programs or activities
- Management support
- Supervisor support

- Employee support (including union support, if applicable)
- Current operational priorities
- Management and employee reward systems
- History of previous change initiatives
- Terminology
- Communication channels

With these issues in mind, the details of the implementation process can be addressed, and should include:

- Steering Team Selection Strategy
- Observer Selection Strategy
- Steering Team Meeting Schedule
- Observation Strategy
    - Whole Site
    - Beta Site
    - Rotating
    - Lone Workers
- Kickoff Strategy
- Communication Strategy
    - Company Newsletter
    - Site Newsletter

- Company Intranet
- Safety Meetings
- Bulletin Boards/Plant T.V.
- Personal Communication Opportunities

On rare occasions, an assessment will discover a site is not ready for BBS. At this point, the decision should be made to either prepare the site or abandon plans for BBS and possibly look for other ways to improve safety. The most common findings to challenge site readiness are:

- Previous failed attempts at BBS
- Little or no traditional safety
- Accidents not behaviorally preventable
- Too many behaviors to address
- Drug and alcohol use
- Resistance and/or lack of support (managers, supervisors, union)
- Upcoming union contract negotiations
- Too many conflicting projects
- Major workplace or workforce changes

If any of these obstacles are discovered, they can usually be addressed in a remediation plan to get the site ready to implement BBS at a future date.

Chapter Questions:

- Does your site accident and near-miss data have adequate detail to analyze it for behavioral accident-prevention opportunities?
- Who is the best person to conduct your assessment and get candid information from site employees?
- What is the best time to conduct the assessment to ensure good participation and minimal disruption of normal operations?
- Did your assessment determine site readiness and the issues that will impact implementation?
- Will BBS potentially help this site move to a better level of performance in safety?
- Can you use an existing safety committee or team for your steering team for BBS?
- If not, who should select your new steering team members?

Part 4: Putting Theory into Action

Chapter 6
Implementing and Customizing Lean Behavior-Based Safety

The customization of the BBS process begins during the last stages of assessment, but continues into the implementation stage. Leaders make strategic decisions after seeing the assessment findings, and the site steering team makes tactical decisions in the customization of the process for the site. This two-staged involvement tends to create the maximum sense of ownership for the process as well as enabling input from the knowledge bases of all levels in the organization.

Site leaders have input in the selection of the steering team members, the alignment of support for the training and implementation, the timetable for implementation and the allocation of resources for the process. The steering team, along with their facilitator or sponsor, have input in the selection of behaviors and overall design of the site checklist, the observation strategy, and their own role in overseeing the process and utilizing the observation data for continuous improvement.

The training and facilitation of the steering team comprises the majority of the implementation process. Additionally, training must take place for observers and briefings should be held for leadership and workers not on the steering team or trained as observers. The steering training is divided into four modules, two delivered prior to process kickoff and two afterwards. Observer training and briefings for leadership and the workforce should occur after the first two modules have been completed by the steering team and the plan for a customized process implementation is designed.

The flow of a process implementation should resemble the following table (post-assessment).

Step	Description	Resource Involved	Timetable	Time Required
ST 1	First Training for Steering Team (BBS Basic Concepts)	Consultant, Steering Team Members, Site Coordinator	When convenient	4-8 hours
ST2	Design Workshop	Consultant, Steering Team Members, Site Coordinator	Together with ST1 if possible	8 hours
M/S	Leadership Training	Consultant, Managers, Supervisors (grouped as available)	After ST2	2 hours per session
OT	Observer Training	Consultant, Observers (grouped as available)	Before Kickoff	4 hours per session
WB	Workforce Briefings	Consultant, Workers in shifts (grouped as available)	Before Kickoff	½ - 1 hour per session

KO	Process Kickoff Events	As Planned, if Needed	As Process is ready to begin	As planned
ST3	Steering Team Training #3 (Data Analysis and Problem Solving)	Consultant, Steering Team Members, Site Coordinator	2-3 months after Kickoff or when data is sufficient	4 hours
ST 4	Steering Team Training #4 (Process Long-term Issues)	Consultant, Steering Team Members, Site Coordinator – Additional Training as Needed	3 months after ST3	4-8 hours
PA 1	Process Audit & Continuous Improvement	Consultant, Strategic Managemen t, Steering Team Members, Site Coordinator – Additional Training as Needed	12 months after Kickoff	8 hours

Steering Team Module 1 - BBS Basics

Since the steering team will customize, manage and continuously improve the BBS process, the successful training of this team is a key to your success. Training

should be customized for the steering team, taking into consideration their education level and the scope of design expected of them after assessment. Team training should consider these topics:

- Basic behavioral science concepts
- How to identify key behaviors to be observed
- How to write behavioral pinpoints
- How to observe and record employee behaviors
- How to track and analyze observation data
- Team problem-solving techniques
- How to reinforce positive behaviors
- Developing and carrying out action plans
- How to strategically plan modifications or improvements to the BBS process
- How to select and train additional individuals for the steering and observation teams
- How to develop a site observation and feedback strategy
- How to monitor and audit the process to keep it on course

As a result of this training, steering team members should have the skills to develop behavioral checklists, observe and record behaviors, provide feedback, and deliver appropriate consequences.

Steering Team Module 2 - Process Design (Customization)

As soon as possible after the Module 1 training, the steering team should begin Module 2, which is a workshop to design and customize the BBS process to reflect the organizational and operational issues at the site. The steering team should be challenged to combine what they just learned in Module 1 with what they already know of the site and its people from their experience to customize the process. The customization and design workshop includes:

1. Branding the process
2. Developing a site-specific checklist of precautions
3. Developing a strategy for observations
4. Defining the flow of observation data and software
5. Defining team structure, meetings and other functions
6. Finalizing the process kickoff strategy
7. Defining the roles, responsibilities and results for everyone involved in operating and supporting the process

The product of this meeting should be a written action plan for the design and implementation of your site-specific Lean BBS process, which includes definitions of mutual roles and responsibilities. The plan should be completely customized for the site needs.

1. Branding the Process

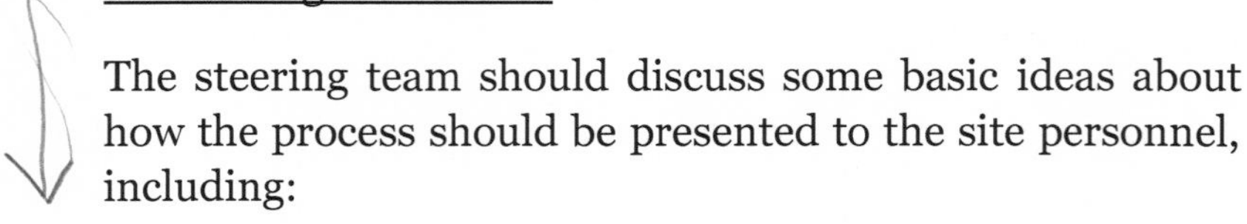

The steering team should discuss some basic ideas about how the process should be presented to the site personnel, including:

- What the process should be called or named
- Whether or not the process should have a visual symbol or logo
- If a logo will be developed, who should be involved
- How and where the logo should be used

Site personnel will label the process if the steering team does not, so many teams elect to control the naming of the process. Logos work well at some sites and are not popular at others. The steering team's familiarity with the site and its people should be relied on to make the right decision. Having contests to name or develop logos is discouraged, as it tends to cause more problems than it solves. Good logos can be used on checklists, communication media, caps, shirts and awards for service to the process if the team decides to use them.

2. Developing a Site-Specific Checklist of Precautions

Selecting the right precautionary behaviors for the site checklist is a crucially important step in a successful Lean BBS process. Working on the wrong behaviors waists energy and resources, and often creates frustration and resistance to future efforts.

The most reliable source of data for selecting precautions is often site accident and/or near-miss data. A Pareto analysis of this data often reveals the behaviors that can have the greatest impact on accidents based on the site's own unique past experience. ProAct Safety has a proprietary worksheet for this analysis but sites often create a generic form of Pareto to discover their highest-impact behaviors.

Many BBS processes focus on at-risk behaviors and direct observations toward stopping these. Such processes create tension and confrontation between observers and workers that sub-optimize results. A better method is to identify the behaviors that can minimize or eliminate risks (precautions) and direct observations to encourage and coach workers to take these precautions. The methods used to stop behaviors tend to damage relationships and culture while the methods to encourage improvement tend to make them stronger.

Checklists in many traditional BBS processes include twenty or more "critical" behaviors. With long checklists, observing and giving feedback can become very time-intensive. Also, long checklists can create dependence on the observations to maintain the consistency of behaviors. When the frequency of observations goes down, the workers tend to quit taking the checklist precautions.

Shorter checklists take less time to observe and gather data, create habitual competence, minimize dependence on ongoing observations, are more easily remembered by workers, and tend to produce quicker and more focused results. They also take a lot less manpower. The Pareto analysis most often identifies 18-25 behaviors that can impact accidents, but usually a few of those behaviors have more impact than the others. Effective checklists limit themselves to a handful or less of behaviors with the greatest potential impact, then adds others after mastering the initial focus.

After determining which behaviors to include on the initial checklist, the steering team should decide which variables will make the data useful for future problem solving. Common variables for checklists include some, but seldom all, of the following:

Observer, Date, Time, Location or Department, Type of Worker Observed (employee, temporary, contractor, etc.), Operating Conditions (normal, planned outage, unplanned outage, etc.), Tenure of Worker Observed, Task (normal job versus cross-trained), Weather Conditions, etc.

The last step of checklist development is the design of the document. Size, weight and color of paper are all options. Early checklists were developed on regular, letter-sized paper and were usually used with clipboards or printed in pads. More recent checklists are often pocket-sized and printed on card stock to make it easy to write on single copies.

Some BBS observation software allows for the use of smart-phone or tablet input of observation data without the need for paper copies. This works well for some sites, but others have conditional issues or simply don't want to bear the additional cost for the devices or software.

3. Developing a Strategy for Observations

Now that the site checklist is developed, the steering team should turn its attention to how the checklist will be utilized. The initial goal should be to observe each worker one time per month (although the goal will likely change in the future). Logistics such as location, shift and movement of workers among locations should be considered in developing the strategy.

The basic elements of observation strategy are:
1. How many observers are needed?
2. How many observations should each observer perform per month?

The range of options goes from one observer observing everyone else at the site to everyone being an observer and observing one other person per month. The ideal solution is almost always somewhere between these two extremes. The underlying issue is a matter of involvement versus quality. Involvement is desirable, but completing only one observation per month will not build expertise and often the quality of observations will suffer. Involving everyone can be accomplished over time rather than all at once and this is often the best option.

Observer Selection - To facilitate change in behavioral performance, observations of employee behaviors must be completed and recorded on a daily basis to measure improvements and to reward successes. Ideally, we would love to have everyone at all levels see value in safety observations, in being able to observe and be observed by anyone regardless of level or job. Who is best to perform observations: everyone or a select few?[3]

Does everyone already have frequent and engaging coaching conversations about observed safe precautions being taken, and expressing concern when perceived at-risk behaviors are witnessed? If not, transitioning to the entire workforce participating as observers will be a new concept to the organization. Never forget, forced change is almost always temporary. When the force goes away, so does the change.

Change strategies should consider trust as the primary factor for the observation process design.

When you mandate something people do not see value in, you are engaging hands and feet without hearts and minds, often creating a false start or sub-optimized results. People may perform observations, but may only do them when they know they have to, or do the minimal necessary to get by. Mandating a process designed to influence, not enforce, rarely yields passionate engagement. In fact, it tends to create disinterest, disengagement and malicious compliance.

If someone does not want to be an observer (e.g., safety coach), but we make them anyway, how will it affect the person being observed? Might their experience influence perceptions about the process? Further, if this experience is negative, might this carry over in how they conduct future observations?

It is generally best to identify your recognized change agents or influential people who see the value in the tool and are willing to participate. Help them show the value in the process by targeting and over communicating early wins. When this occurs, people will respond and start pulling toward a process they see offers the company and, most importantly them, value. If your culture is indeed ready for all to perform, it could compromise success if only the "chosen" few are involved.

Starting with too much or not enough participation could stall or expedite results. Either way, it's vital to understand what motivates, or could demotivate, your culture with a BBS process. Generally, it is better to grow the process into the culture that creates pull rather than pushing it into place. Work on growing participation based on value recognition and interest that creates want-to rather than have-to engagement. Most importantly, and considering acceptance to change, make the strategy fit your culture rather than making the culture fit the strategy.

Observer Training - The techniques used to complete behavioral observations will impact the results of your BBS process. Therefore, you need thorough training of observers to help them build the skills they need to complete and record observations effectively. The observer coaching process (explained later) allows your organization to develop internal resources to maintain the process after the initial implementation phase. Observer training typically lasts four to six hours and focuses on:

- Basics of behavioral science concepts
- Your observation process as designed by your steering team
- Techniques for observing and recording observations
- Providing coaching feedback to employees
- Analyzing and posting observation data

Evolving Observation Strategy - Don't let your steering team get stuck on the initial observation strategy, because it needs to evolve over time. It takes more reminders and coaching to change a habit than to maintain it after it is changed.

At a recent conference, the number one problem reported by mature behavior-based safety process teams was observer burnout. Why do processes falter and why do observers burn out? The answer to both of these issues lies in the strategy for observations.

Most BBS teams were taught an observation strategy that simply blanketed observations evenly across the site. They quickly learned that there was a direct correlation between the number of observations and the impact on accident reduction. The goal of the process became to hit the target number of observations. As the process matured, it increasingly became more difficult to accomplish the goal. Additionally, hitting the same number of observations began to have a diminishing impact on accident rates. But, since the process was successful, it continued.

Leaders of mature BBS processes who revised the strategy for observations have recognized additional gains and created a more resource-efficient, sustainable approach. The challenge is to modify the perceptual goal of BBS from the quantity to the quality of observations.

What triggers an observation? When we ask, "Why do you perform observations?" we often hear answers like: "Because we have a numbers goal," "Because we have to," and "Because if I don't, I'll get in trouble." The goal of a behavioral approach to safety should not be to simply accomplish observations. The goal should be to improve safety by providing simple strategies that can be easily internalized and by identifying influences on at-risk performance.

People do things for a reason. If you want to improve performance, you need to first identify what is influencing current performance. Is it a perception, habit, obstacle or barrier?[4]

If you can eliminate or mitigate the influence on risk, you are enabling a sustainable behavioral change. Observations that do not gather insight into the reasons for risk are functioning solely as antecedents, activators or triggers that remind and refocus people on certain precautions. This is the most costly activator you can use. What triggers an observation should be used in strategic response to previous observations, not simply a numbers goal.

Common Alternative Observation Strategies - Four advanced observation strategies have been identified which have resulted in significant returns with little internal investment.

- **Self-Observations** – This approach is ideal for isolated workers or lone drivers and where traditional observations are not an option. It can also be used to supplement traditional observations and further reinforce a change strategy. However, there are limitations to self-observations that must be considered.

  Workers are often blind to their own habits, so self-observations need to be supplemented by outside observations, at least periodically. Also, reflection can be grossly inaccurate, so workers cannot simply ask themselves, "How did I do today?" Self-observations need a reminder mechanism to indicate when to start and stop, and this often requires some innovative thinking to design.

- **Aimed Observations** – Many organizations analyzing their observation data identify visible trends in risk, such as time of day, day of week, task, weather, etc. The observations can be aimed at these exposure targets, rather than blanketed. If insight into influences (asking why) is not collected during observations, the ability to address risk exposure is limited.

- **Blitz Observations** – Like aimed observations, a blitz is a focus on a target. But, rather than asking observers to aim individual observations at a target, blitzes send groups of observers to the targets to do multiple, simultaneous observations. Blitzes concentrate a lot of attention and can result in quick improvements.
- **SWEEP Observations** – SWEEP is an acronym that stands for "Seeing Without Explaining to Every Person". SWEEP observations cannot be used alone to accomplish behavioral change since there is no feedback component. SWEEPs are simply an ongoing way to aim other types of observations at the targets where they can do the most good. Some organizations use SWEEPs to determine a more accurate percent-safe. Other types of observations give workers notice before the observation begins, which often results in an artificially high percent-safe. Caution: SWEEP observations should only be used when the culture and employees have developed a high level of trust in the BBS process.

Careful Consideration to Progress - Mature behavior-based safety processes often are doing the right things. Further improvement lies in doing those things better. This same principle will apply to how you choose to help your BBS team improve their process. Involve the steering team and the union, if applicable, in selecting the new strategies.[5]

Prior to implementing new processes or changing existing ones, it is critical to discuss with the workforce the purpose and details of the new observation strategy. A different observation strategy can breathe new life and energy into an old BBS process.

Coaching Observations - A coaching observation is when one observer performs an observation of another observer as he/she performs an observation. As with anything, poor habits are developed when conducting safety observations and they will go unnoticed unless someone else points them out. A coaching observation is intended to be positive and helpful. The steering team should be primarily responsible for developing a schedule or plan for conducting these coaching events.

The observer conducting the coaching should use an AWARE checklist, ask "Why?" and give feedback as if a regular safety observation was being completed.

A.W.A.R.E. Observer Coaching Checklist

Observer Name:	**Coach Name:**		**Date:**	
Observation Item	**√**	**Concern**	**Reason for Concern What/Why**	
Announce				
Watch				
Ask				
Reinforce				
Express Concern				

4. Defining the Flow of Observation Data and Software

There are several software packages and services available that are designed to analyze and report BBS observation data. Many sites and companies develop their own databases or spreadsheets to accomplish this as well. The main consideration for the steering team is to make their checklist as compatible with the software as possible.

Unless you choose to use direct input of observation data via smartphone or tablet, you will have paper for observers to fill out. The steering team needs to decide what observers should do with completed observation sheets, how they should be entered into the software, and how the data should be made available to the team at their monthly meetings.

Confidentiality of the data should be a major consideration. Many teams designate data input personnel and data drop-off sites or install lock boxes for data collection. Some teams have each observer input their own data if all observers have computer access. After data is entered, most teams shred or otherwise destroy the paper copies to maintain confidentiality.

Steering team meetings should be timed so the data from the previous month has been entered completely by the time the team reviews it. Ideally, the team can meet where data access is available and can be referred to as needed. If not, copies of data reports should be made for each team member.

5. Defining Team Structure, Meetings and Other Functions

The steering team should decide on their own leadership. Many teams have a self-directed style and simply share responsibilities without any formal leadership structure. Others have traditional or designated team or committee structures with leadership defined. If a team uses a structured style with defined leadership, it is often wise to make the term of leadership short so it can be changed without long-term impact on the team.

The team should also decide on where, when and how frequently they should meet. Teams often meet every week or two for the first 1-3 months after kickoff to get the observation process started. Monthly meetings are usually sufficient beyond that. The meeting location should be convenient for all team members and ideally have computer access to observation data. The times for the meeting should be considerate of the work schedules of the team members. If any of these facets of team function are found to be awkward or difficult, the team can always try alternatives.

6. Finalizing the Process Kickoff Strategy

Kickoff activities fall into two categories: required and optional. These activities should be scheduled and completed just after Steering Training #2 since that's where many details of the process will be decided.

The required activities are:

- Leadership must be briefed on the steering team's plan for implementation, and their support roles in that plan must be defined and committed to.

- Observers must be trained and given their start-up assignments for who and/or where they are to observe and how often.

- The workforce must be briefed on the details of the process so reasonable expectations are created and questions answered prior to kickoff of observations.

  (It may take several training sessions to complete these required activities depending on the size of the site workforce and availability of workers around normal work routines and shifts.)

Optional kickoff activities might include the following:

- Meetings to formally kickoff the process
- Celebrations of the new era of safety about to begin
- Meals
- Announcements
- Promotional items
- Sites with a history or tradition of starting new projects in a certain way would be wise to continue those practices with Lean BBS. If there is no historical or traditional reason for such events, the steering team should simply decide if such activities or events would add value to the process and if that value is worth the costs of doing them in terms of time and resources.

7. Defining the Roles, Responsibilities and Results for Everyone Involved in Operating and Supporting the Process

Lean BBS is most successful when everyone knows and does their part. Every person at the site will play a role, have specific responsibilities, and be expected to produce certain results to make the process work. The examples that follow are generic RRRs (Roles, Responsibilities and Results) for the common levels of people in most organizations. These can be used as-is or modified to fit the site culture.

These RRRs should be agreed upon by the steering team and their sponsor or facilitator and be a part of training for observers, leadership and workers. Many teams print copies for everyone and post them in public places for regular reference and review.

Sample RRRs for Site Manager/Management

Roles

- Selector of Steering Committee Members
- Support Accountability Manager
- Resource for Steering Committee Action Plans
- Cheerleader for BBS
- o Confidence Builder
- o Motivator

Responsibilities

- Provide Resources for BBS Success:
    - Consultant Assistance
    - Time
    - Training
    - Meeting Space
    - Checklist Supplies
    - Software
    - Communication Media
    - Expert Advice
- Hold leadership accountable for BBS support
- Approve or reject action plan requests for help from the steering team
- Refrain from using BBS data in any manner that is negative
- Do not discipline or reprimand employees for data generated by the BBS process
- Try not to focus on areas of the BBS process that may not be meeting all targets or goals; instead, focus on what the committee has accomplished and this will encourage them to meet their goals

- Approve "easy win" projects early on in the implementation of the process to show support and add motivation to the steering team
- Commit to approving safety-related projects determined a priority by the steering team and validated by gathered data
- Show enthusiasm and optimistic support of BBS and the process
- Inquire about the BBS process often and give specific positive reinforcement whenever possible
- Allow the steering team to have a budget to support their process
- Ensure supervisors understand the commitment to this process and know their role in supporting it.

Results

- Steering team members will feel valued and supported
- BBS will have needed time and resources (priority)
- Successes in BBS will be recognized and celebrated
- Steering team action plans will be implemented or rejected with explanation
- Support for BBS will be consistent and improving

Sample RRRs for Supervisors

Roles

- Facilitator of the Observation Process
- Scheduler of Coverage for Steering Team Members and Observers
- Encourager of BBS Participation

Responsibilities

- Set a tone of cooperation with BBS participants
- Communicate the role, priority and importance of BBS to all employees
- Work out scheduling issues for steering team members and observers
- Encourage workers to participate and cooperate with BBS

Results

- BBS will operate smoothly in your area
- BBS participants will feel accommodated and appreciated
- Suggestions from the BBS steering team will be implemented or explained why not

Sample RRRs for Site Facilitator/Sponsor

Roles

- Site BBS SME
- Advisor to the BBS Steering Committee
- Provider of Resources to the BBS Process
- Communication Link Between BBS and Site Leaders
- Facilitator of Steering Team Meetings

Responsibilities

- Learn and keep up-to-date on BBS information and strategies
- Meet with the steering team
- Advise the steering team
- Facilitate steering team meetings
- Communicate BBS progress to managers
- Accept assignments from the steering team for action items when needed and appropriate
- Assist steering team in developing process indicator metrics that integrate with other facility metrics

Results

- Steering team meetings will be regular and productive
- Steering team needs will be met
- BBS progress will be known by managers
- The BBS process will be successful at the site

Sample RRRs for Steering Team Members

Roles

- Employee Subject Matter Expert in BBS
- Designer of BBS for the Site
- Director of BBS for the Site
- Promoter of BBS to Fellow Employees

Responsibilities

- Become and remain a BBS SME
- Design and continuously improve:
    - Checklist
    - Observation strategy
    - Data management strategy
    - Succession planning
    - Motivation strategies
    - Professional development for the steering team
- Meet monthly, or as planned, to:
    - Review observation data
    - Develop action plans to improve the process and results
    - Review progress on past action plans
    - Plan and execute communication strategies
- Meet with observers to solve problems and improve observation quality
- Audit and adjust the BBS process
- Communicate and publicize:
    - Process issues
    - Results
    - Contributions and contributors

Results

- BBS will produce great KPIs
- BBS will produce great accident reductions
- BBS will sustain long term

Sample RRRs for Observers

Roles

- Coach Safety Improvement
- Set a Personal Example for Safety
- Support the Steering Team

Responsibilities

- CREW (Constantly Remind Every Worker)
- Gather insight on risks and reasons
- Look for checklist items and issues
- Look for other items and issues
- Give feedback to observed workers (positive reinforcement and concerns)
- Exemplify safety compliance and culture
- Identify barriers and obstacles to safe behavior
- Prepare to become a steering team member, if needed

Results

- Every worker assigned to you will be constantly:
    - Reminded to work safely
    - Refocused on checklist behaviors
    - Challenged to question perceptions and habits
    - Asked to identify barriers and obstacles
    - Listened to for questions and suggestions

Sample RRRs for Workers

Roles

- Set a Personal Example of Safety Excellence
- Analyst of Job-specific Safety Issues
- Collaborator on Solving Safety Problems

Responsibilities

- Cooperate with the observation process
- Be open to opportunities to improve safety-related behaviors
- Become a better self-checker for workplace risks
- Discuss safety issues with observers and offer suggestions
- Help observers to identify barriers and obstacles to safety
- Set a good example for your fellow workers

Results

- Constantly increase safety awareness and analysis
- Focus on checklist items to improve safety
- Cooperate with the BBS process
- Volunteer to participate if possible

Leadership Training

As soon as possible after completing Steering Team Module 2, key site managers should be briefed on the plans for implementation, trained in the basics of the process, and fully understand their RRRs for the process. Their support for the implementation plan should be obtained or whatever modification to the plan made. If there are too many managers to get into one session, the main ones should finalize or approve the implementation plan before other managers (and supervisors) are trained and briefed.

It can be critical in some organizations to have the right person(s) to conduct the leadership training. This should be someone who is either an authority or trusted by the organizational leaders, and should have sufficient knowledge of BBS to answer potentially higher-level questions.

These briefings should accomplish three basic objectives: 1. Give attendees a good understanding of the philosophy of BBS and the process steps, 2. Give attendees details of the steering team's implementation plan for the site, and 3. Define the RRRs for the attendees and solicit their commitment to support the process.

When the leaders don't lead, the followers don't follow. This is especially true in safety. Site leaders constantly communicate priorities and strategies to their workers, whether they intend to or not. With training, leaders can take active control of the messages they send to promote safety as an organizational value rather than a changing priority. They can set levels of expectation to point everyone in the direction of safety excellence and exert a positive influence on the formation of safety culture.

This training should be tailored to your organization to build true support and direction at the top for whatever programs and processes are working with the rank and file. Site leaders will quickly learn techniques to increase their effectiveness in influencing others and directing the efforts of the organization toward worthwhile goals. Effective leadership is critical to creating and supporting a true culture of safety awareness. Leaders must understand how and why this process works and actively use positive reinforcement techniques to promote and reinforce desired employee behaviors.

Observer Training

If the steering team has not been trained on how to complete observations, they should also attend observer training.

Observer training has four basic parts: 1. Understanding influences on risks, 2. How to perform an observation 3. Classroom and workplace practice in performing observations with on-the-job coaching, and 4. A review of observer RRRs and a commitment to perform as trained.

1. Understanding Influences on Risks

People take risks for a reason. If we don't know the reason, we may not be able to change the risk-taking behavior. When we examine the complexities of most work environments, we find worker behavior is influenced by a number of factors. People do not make decisions in a vacuum. Their thinking, surroundings, past consequences and cultural reinforcements play a critical part in their decisions. Peer observation and feedback is just one example of possible influences. One of the most effective and sustainable strategies to improve safety involves understanding, measuring and managing the influences on risk taking.

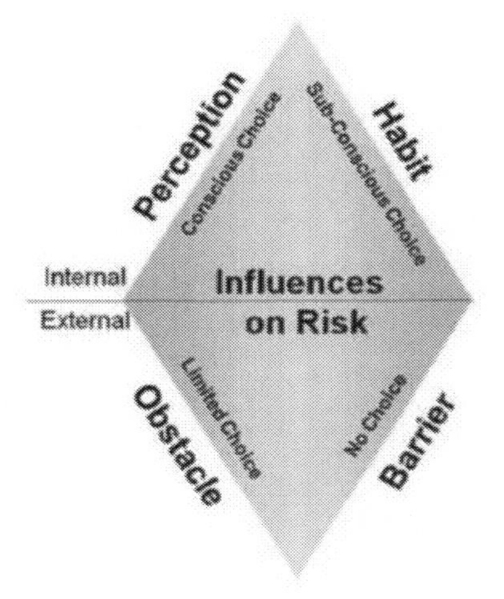

The Risk-Taking Influences Model – This simple four-part model has been used by hundreds of organizations to effectively improve safety. It has proven powerful due to its simplicity, practicality and ease of use compared to other, more complex models of risk causation and human error classification. Most importantly, it is action-oriented and prescriptive. Once a reason for risk is classified and quantified, a solution method can be quickly identified and proactively implemented.

When this model is put into practice, it quickly becomes evident that if one worker is influenced to take a risk (consciously or subconsciously), others are as well. Proactively understanding and addressing the influences on risk at the cultural level provides superior opportunities for sustainability of safe group behavior and reduction of exposure to risk when compared to the traditional approach of addressing individual behavior.

The model also helps to classify the reason for the risk by categorizing worker's answers to the "why" question. Here is how the model works: When a supervisor or peer sees a potential risk, they express concern to the worker(s) involved and gather insight by asking about the rationale for performing a task in a certain manner. This discovery-learning conversation, more often than not, provides upstream safety indicators in the form of influences on risk.

The response to the question provides profound knowledge which is classified into one of four influence categories. The four categories of response are perception, habit, obstacle and barrier.

INFLUENCE 1 - PERCEPTION

Hazard recognition is not just about seeing a hazard. It is about mentally acknowledging "it could happen to me." Many workers who have been injured had the knowledge of the potential hazard, but lacked a sense of vulnerability. This is often due to the inability to identify the potential consequences of low-probability risks.

Unfortunately, the common measurements in safety tend to perpetuate this undesirable mentality (i.e., zero accidents means we are safe; it is safe as long as you do not get injured, etc.). If someone frequently takes a risk but remains injury-free, the perception of risk tends to become skewed. When the rationale for the risk is questioned, the response one will hear will be similar to one or more of the following:

1. In my opinion....
2. In my experience....

3. I don't think it's a problem because....

4. I've done it before and not gotten hurt.

5. What's wrong with it?

INFLUENCE 2 - HABIT

While "habit" is not a scientific term, it is used in common language and is typically easy to understand when explaining the model. Dictionary.com defines a habit as "an acquired behavior pattern regularly followed until it has become almost involuntary." A person will make a decision to perform a task in a particular manner until subconscious mental templates are created. Once enough experience has been obtained, the individual no longer has to consciously think about the task.

When is the last time you had to consciously think about opening a door? You have developed a template for opening doors and the thinking now occurs subconsciously. This is similar to how a work habit is formed.

Habits, like perceptions, are within the control of the worker to change. However, they are not changed in the same manner. Information will often change a perception; it will not necessarily change a habit. Habits are deeply engrained and rooted into the patterns of human behavior and require repeated antecedents or reminders. The required frequency and spacing of these reminders will differ depending on the complexity of the task and the length of time the habit has been formed.

When questioned about the rationale for a risk, a habitual response will often sound like:

1. That's the way I always do it.
2. I don't know.
3. I don't think about it.
4. It's the way we do it around here.

INFLUENCE 3 - OBSTACLE

An obstacle is something that makes it difficult to perform a task safely or take a safety precaution. Obstacles are often inadvertently created during design or modification of work stations or operational processes. Common obstacles include inconvenient location or difficult access to tools and equipment. The removal or reduction of obstacles, most often, is not within the control of workers.

When questioned about the rationale for a risk, a response to an obstacle will often sound like:

1. It would be difficult to do it that way because....
2. If I do it that way, (this would happen).
3. It takes too long to do it that way because...

INFLUENCE 4 - BARRIER

A barrier makes it impossible to perform a task safely or take a precaution. It is best for barriers to be responded to first, as they negate the ability to perform the task safely or take safety precautions that control exposure to risk. Complete unavailability of correct tools or equipment, or lack of anchors or tie-off points for fall protection are examples of barriers.

When questioned about the rationale for a risk, a response to a barrier will often sound like:

1. There is no way....
2. It is impossible because of....
3. I can't because of....

Responding to Determined Influences - Once the influence has been identified and categorized, appropriate action can be taken to address it.

For perception influences, distribute information about the risk through training, articles/talks, in safety meetings and help from leadership. Stress the importance of the particular risk in accident reports, add risk avoidance techniques to new-employee orientation and discuss risks and risk avoidance in pre-job briefings and shift huddles.

For habit influences, increase the number and frequency of reminders in target areas through signage to prompt action, behavioral observations, reminders in safety and other meetings, creating a buddy system to have workers reinforce each other and distributing pocket cards for reference.

For obstacle and barrier influences, first identify the issue from accident and observation data to determine exposure rate, then define the potential impact of inaction and ask for help in removing the obstacle or barrier from management or engineering and, finally, apply the hierarchy of control model.

This holistic model has been successful due to its positive impact on culture, systems and individual actions. The traditional approach to "correcting individuals" often only solves a problem short-term. Identifying and modifying influences produces long-term results that span generations of employees and managers.

When leaders first ask "why" to gather insight, rather than quickly blaming or confronting, it creates an ongoing safety forum between management and employees. In doing so, a climate is created in which discussions about the problems and opportunities to improve safety are more culturally acceptable. Additionally, expressing concern and asking for the rationale behind a decision has proven to be a more effective method to change individual behavior. Direct confrontation often can cause a defensive response, or worse, conceal the underlying reason for the risk.

2. How to Perform an Observation

Effective observations are all about being A.W.A.R.E.SM A good observation begins with the site checklist with the most impactful precautions to prevent accidents. Observers often see risks workers are not aware of and offer feedback to make them more aware. When observers ask workers why they did or did not take precautions, they become aware of the influences on risks. The steps of a good observation are about being A.W.A.R.E., and this acronym helps observers remember the 5 steps of a good observation:

Announce - The first step in the observation process is to let the person you intend to observe know you are there. This initial contact sets the tone for the observation and resulting discussion. It is normal to wonder, "If they know I am there, they might do everything right." Wouldn't this be desirable? One of the goals should be to help create new habits, not to catch a rule violator. Another step of the methodology will reinforce this. However, if someone feels ambushed or spied on, how might trust be compromised? Make sure everyone you will observe, and even those nearby, is aware of the observation before you begin.

Watch - After the individual is aware, you should spend your designated time watching the job task. What will you look for? Ideally the answer is, "Can the employee perform the task safely?" and "Do I see anything that concerns me?" Lean Behavior-Based Safety processes identify what safe looks like by positively defining the significant few precautions employees can take to reduce the probability of incidents. Generally, this is better received than an observation attempting to determine if rules are being abided. The purpose of observations should be to identify if workers can perform the task safely and proactively identify concerns that might increase the chances of an injury. This should not be used as a faultfinding, gotcha or catch-the-rule-breaker opportunity.

Ask - One of the most important aspects of an observation is determining why a precaution was or was not taken. This insight is one of the most effective mechanisms to affect behavior change and prioritize safety improvement initiatives. It is easy for all of us to become complacent with a task often performed. It is important the individual being observed recognizes the rationale for the decisions he is making, for both the ones that reduce risk exposure and the ones that introduce risk exposure. If you see a safe precaution being taken, or an exposure to risk, ask the most appropriate questions: Why did you do it that way? Is that the way you always do it? Do you feel safe doing it that way? Is there a safer way to do it? Were you trained to do it that way?

Reinforce - Observations are an opportunity to specifically point out the positive things a person is doing for his own safety. Emphasis should be placed on reinforcing what the worker is doing right to ensure he is not just being lucky when it comes to injury prevention. If an individual has performed a discretionary precaution while performing his work, this is an excellent time to reinforce precisely what you observed him doing, and encourage him to continue. This helps change the common belief safe is defined by the lack of accidents rather than by what people do to control risk exposure.

Express Concern - When risk is identified during an observation, the language chosen to provide feedback is critical. Expressing concern is a preferred approach over stating someone is "at risk" or "unsafe." When you choose the latter examples, your opinions are introduced into the conversation, often compromising trust and respect. If an observer states concern with how a task is performed, this offers a better chance for a conversation leading to an understanding of why risk is a part of the task.

Teaching observers the A.W.A.R.E. steps of an effective observation and feedback methodology help them remember the important parts of an observation and not skip anything that could improve safety.

3. Classroom and Workplace Practice in Performing Observations with On-the-job Coaching

In the classroom, observer trainees can be given scenarios or watch example videos to practice seeing and recognizing the targeted items on the checklist as they might occur in the workplace. They can also practice giving and receiving feedback on both safe and concerned behaviors. It is important training not just be an information dump, but an application of the process.

In the workplace, new observers can go out in groups and/or with trainers to do practice observations. Carefully choose the workers to observe so as not to give new observers challenges they are not ready to handle. Debrief after the practice session and make sure all questions are answered and the new observers have gained confidence as well as competence from the session.

4. Review of Observer RRRs and a Commitment to Perform as Trained

The final part of observer training should be a review of the observer's RRRs to make sure there is a good understanding of what is expected. Observers should be asked to commit to perform the observations as they have been trained to do. Observers who state their intent to perform are much more likely to perform good observations.

This is also an opportunity to give each observer their assignment for who to observe and how often if that is already determined. Many sites pair up observers with trainers or steering team members for their first several observations before sending them out on their own.

Worker Training & Briefing

The remaining employees participate in a half-hour orientation to the process and are only exposed to the more extensive observer training when, and if, they take on the role of being an observer. This training can be delivered to the workers in shifts or all at once.

It is not necessary that each worker be as thoroughly trained as steering team members or observers, but it is critical they know enough about the process to not feel ambushed when the observations begin. Worker training usually includes some or all of the following items:

- A history of why the site chose to implement BBS
- Introduction of the site steering team
- A list of observers and their assigned areas
- A copy of the site checklist
- An overview of BBS basics
- Details of how the steering team customized the process for the site
- An example of how an observation will be conducted
- Review of worker RRRs
- Question and answer session

It is important to keep worker briefings short and to-the-point since they are usually held just before or after work shifts and attention span is limited.

Kickoff

Kickoff events and publicity are optional and should be considered based on what value they could potentially add to the process. It might be important to hold kickoff events or meetings if doing so is customary in the organization. If every important process has had a big kickoff, not giving one to BBS might send the message that it's not really important.

However, if such events are not customary, it is questionable if they would add value. Discuss the possibilities and decide on the best course of action, realizing expenditures for events might need management approval. Some examples of activities to promote the new process include: meetings, celebrations, meals, announcements and/or promotional items.

Publicity is also optional but, if chosen, the timing should correspond to the beginning of the observation process.

- Internal Media
  - Newsletter
  - Meetings
  - Bulletin Boards/Banners
- External Media
  - Local Newspapers
  - Local Television Stations
  - Local Radio Stations
  - Community Organizations

Steering Team Module 3 – Steering BBS with Observation Data

A key success factor in any implemented BBS process is data management. The data is what enables continuous improvement and helps the observers keep score. Sometimes the reason workers are not improving in safety is because something is getting in the way. Identification of safety barriers and measuring their impact is a powerful tool in improving safety. Traditional safety tends to only focus on lagging indicators and failure rates. Percent safe is a great leading metric for comparison to the downstream metrics of accident rates, severity rates, costs of accidents, etc.

This session should assist the steering team with the steps of designing a good data management and problem-solving program in conjunction with their behavior-based safety efforts. This data flows to the steering team and helps them to remove barriers to safety and change influences that could tempt workers to take risks. Additionally, the ability to isolate problem areas increases your ability to focus corrective effort, reducing wasted resources. The observations are most definitely a great tool for beginning the creation of a culture of safety awareness and development of a personal safety focus; however, without a good BBS data management strategy, the process will most likely not be sustainable. A similar example is a black hole work order system, where you put in suggestions but nothing ever comes out of it.

The data usually tells you its own weaknesses (i.e. too little, not representative, not complete, 'what's instead of 'why's on comments, etc.). Once you adjust the data gathering process, it starts to tell you where your greatest risks are and why workers are taking them. The observation data combined with the original Pareto analysis data helps to prioritize the risk issues for the team to address.

This session also teaches how to look for trends in performance. It is important to know if risk taking is increasing, decreasing or remaining relatively constant. Since observers cannot see every behavior at the site, it is important the behaviors sampled are representative of what is happening across all times and locations. If data is bunched into certain times or locations, it may not be reliable.

BBS is only as effective as the data it generates and uses. The following questions can create or fine-tune observation and data-analysis strategies, and lay the foundation for the four types of effective and efficient action plans.

Is Your Data Quantity Adequate? Are you regularly meeting your goal for desired number of people observed? If so, move to the next question. If not, enact Action Plan 1: Increase the number of observations. Some strategies to implement this action plan could be sharing data with observers to help them understand its use and importance, reviewing your observation strategy (number of observers and observations, etc.), using back-up or reserve observers to cover when regular observers are unable, setting reminders for observers, retraining observers who are not accomplishing their target numbers of observations, stressing the importance of adequate sample for trends and why this process is so important, posting data on targets versus actual observations, pairing observers into teams to support and encourage each other, utilizing blitz observations, creating regular reminders of commitment and performance, and rewarding and/or celebrating when targets are reached.

Is Your Data Quality Adequate? If you have *enough* data, how is the quality? Is the checklist complete with all the variables, including the reasons why precautionary behaviors are not being performed? If so, move to the next question. If not, enact Action Plan 2**:** Improve the quality of observation data. Some strategies to implement this action plan could be giving regular feedback to observers on the quality of data, partnering observers with team members to help improve quality, partnering observers who are having difficulty with stronger-performing observers, training or retraining on observation data details and methodology, reviewing data in observer meetings, showing examples of what a complete well-filled-out checklist looks like, and adding checkboxes or categories to checklists as prompts.

Can You Prioritize and Select a Behavior? Are you already focusing on a single or small group of behaviors to make the biggest impact based on observation data? If so, move to the next question. If not, look at secondary criteria (time of day, day of week, tenure of person observed, shift, department, weather, etc.) to drive focus for improvement.

Can You Identify Influence? If not, enact Action Plan 3**:** Gather better data on reason for concern. Some strategies to implement this action plan could be asking observers to focus on the targeted behavior, looking for the influencer everywhere, asking about it even when the behavior isn't observed, targeting areas or jobs where you have historically had the most of this type of incident, asking everyone for ideas to solve the problem, and involving workers and observers as problem solvers.

If the influence can be identified, enact Action Plan 4: Address influences on risks. Some strategies to implement this action plan for *perception influences* could be training, reading articles or toolbox talks in safety meetings, helping managers and supervisors reinforce and coach, stressing the influence in incident reports, and adding the influence to new-employee orientation. For *habitual influences*, some strategies could be increasing the number and frequency of observations in target areas, placing signage about the behavior, and having reminders in safety and other meetings. For *obstacle and barrier influences*, some strategies could be, identifying the issue from incident and observation data, defining the potential impact of inaction, and asking for help from those with a larger budget.

Consider leveraging this methodology and ensure your data collection and analysis strategy focuses on efficiency and capturing and delivering value to the customers of the process.

The main topics to be covered in this module are:

- Analyzing data to understand what it tells you
- Prioritizing issues to address first
- Understanding the scope of safety improvement efforts
- Developing action plans to address perceptions and habits
- Asking for help when barriers or obstacles to safe behaviors are discovered

- What data should be posted, shared or sent to management (if any)
- Communicating to the workforce what action plans are in place and how workers can help
- Sharing successes with the workforce when action plans produce positive results

Steering Team Module 4 – Navigating to Maintain Process Viability

At this point in the process, all the steps of the Lean BBS process are in place. The purpose of this final module is to keep the process viable long-term. This module consists of six steps the steering team should review periodically to make course corrections, if needed, to the process. The six steps are:

1. Process Audits – Periodically, the steering team should audit the process to make sure they are working their plan and their plan is working. They should check to see if they are getting good participation and support, are hitting performance targets such as number of observations, the behaviors on the checklist are actually improving and becoming more regular, and the improved behaviors are impacting the safety lagging indicators.
2. Motivation – The steering team should have a plan to motivate the observers and workers, and managers should have a plan to motivate the steering team. The plans need not be expensive or complicated but should show a true appreciation for participation in the process.

3. Succession Planning – Steering team members and observers should be challenged to identify and prepare their own replacements over time, realizing none of them will be there and able to participate forever. The best time to find replacements is *before* you need them.

4. Follow-up Training – At the time of kickoff, everyone will have been trained in BBS basics and be familiar with their RRRs. How will the site train new employees as normal attrition and turnover occur? How will BBS become part of new-employee orientation and who would train a new site manager if you get one? Who will train new observers as they rotate? The team should develop and execute a plan.

5. Professional Development – The steering team will be busy for the first year or more implementing the process. After that, they will need a source for ongoing professional development and new ideas for their process. They should consult with management to see what resources are available (subscribe to safety publications, network with other sites, or send team members to conferences). They should first make use of all free information and resources available on the internet, but other sources can be valuable to ongoing improvements.

6. Maintaining a Results Orientation – The steering team must periodically ask themselves why they continue Lean BBS. If the answer is simply to get more observations or hold meetings, they should recognize they are losing their results orientation. It is necessary to work the process if you expect results, but working the process is just a means to an end. The goal of Lean BBS is to improve safety and reduce accidents. If the process

is not doing that, it should be changed and redirected. Teams must not lose sight of the end goal.

Chapter Questions:

- Do you have a clear perception of what it would take to implement Lean BBS at your site?
- Do you have adequate support for, and understanding of, BBS at the management and supervisory level?
- What resources do you have to deliver training?
- Who would be a good sponsor or facilitator for your steering team?
- How would you envision observation data being input and reported at your site?
- What is the best time of the year to begin your implementation?

Chapter 7

Auditing the Lean BBS Process

The ultimate goal of Lean BBS is that it become a self-auditing process. If the steering team feels confident in self-auditing, they should be allowed to do so with the oversight of the sponsor or consultant. However, the establishment of self-auditing may require a jump start. The internal or external consultant can audit the process to get this started, but care should be taken to encourage as much participation from the steering team as possible. The best approach is often for the auditor to ask rather than tell. This applies both to discovery of current status as well as steps to address deficits or further improve the process.

The initial process audit should be conducted approximately one year after observations begin. If the process is struggling, it is okay to audit it sooner; and if it is operating smoothly, it can be pushed back a month or so. However, avoid delaying too long as problems can arise or become more acute during this period. The audit should take no more than half a day, with another half-day for corrective action or adjustments, if necessary. All steering team members should be in attendance for either a self-audit or outside audit to maintain ownership of the process.

The focus areas of the audit are each part of the Lean BBS process:

Steering Team – Assess whether the team has the right number and types of members. Is it functioning efficiently and are meetings well attended? Is the team managing the observation and data-utilization processes? Are they burning out? Have they had turnover? Do new members need additional training? How do they feel about working together?

Observation Process – Assess if observers know and perform their jobs, and if observations are perceived as helpful by workers. Can observers name the steps of an observation? Are they hitting their target number of observations regularly? Do they feel they are helping workers form habits around the checklist behaviors and identifying influences that might discourage safe behaviors? Are they aware of action plans initiated by the steering team?

Observation Data Utilization – Assess if the steering team is gathering and using good observation data to formulate action plans to facilitate safe behaviors in the workplace. Does the data form a good cross sample of workplace activity? Does it include influences that impact worker's behavioral decisions on the job? Are action plans well communicated to observers and workers?

Results Orientation – Assess if the steering team is directing their efforts toward desired results and not just process indicators. To stay results-focused, there are five indicators of a behavior-based safety process that must be managed. Consider how to leverage the model outlined below to improve efforts with your process.[6]

Measurement 1: % of Observation Target – Most organizations have a goal to observe and provide feedback to a set number of people in their operations on a monthly or weekly basis. This goal changes based on the culture and process maturity, and is typically set by number of people, rather than number of activities.

Measurement 2: % Action Plan Tracking – Like observations, measuring the number of created and closed action plans could prompt an action plan to create action plans. Most clients tend to measure the number of closed versus completed, with the intent to not create many, but to create the right ones.

Measurement 3: % Participation – How many trained individuals are reaching their target number of observations?

Measurement 4: % Knowledge – There are two indicators that provide insight into knowledge. 1.) How many people can recite from memory the precautions you are focusing on? If it never gets in their memory, it will never get in their habit. 2.) How many people can name a recent success, or three? If the customers of this process (everyone) do not view it as effective or valuable, support will wane.

Measurement 5: % Safe – Whether you are focusing on one behavior or many, what is the average percent safe during your observations for the focus behaviors?

A Single BBS Leading Indicator: Add the totals of all measurements and divide by the total (5). This provides you a single BBS indicator that outlines the health of a process. Moreover, this information provides valuable insight to those whose support is necessary for continued success or investment in responding to action plans.

Of course, the qualifying metrics are lagging indicators which help manage the focus within the BBS process. If the process is completing observations, maintaining great participation, creating action plans, and knowledge is high but incidents are still occurring, perhaps it is focusing on the wrong behavior, time, area, tenure of employee, etc. It is possible to successfully work your process and be working on the wrong things. Alternatively, if you are completing observations but not action plans, participation is low at meetings and no one knows the focus or successes, you may have good results due simply to luck.

Any reported or activity-based measurement can be manipulated to meet goals, so be careful about setting number requirements. The goal of this tool is not to provide more accountability opportunities to focus people on numbers; rather, it is to better understand the health of your process and to feel more confident in where to focus time, energy and resource decisions.

Remember, any safety measurement system is only as effective as the integrity of and belief in the tool, the priority placed on it by leadership, the value recognized by its customers, and internal capability to make methodology and systems adjustments based on the measurement outcomes.

Management Support – Assess if the steering team members feel they have adequate resources and reinforcement from site managers. Do they have meeting time and space? Do they have resources for checklist and data entry activities? Do managers say positive things about the members and the process?

Supervisory Cooperation – Assess if supervisors enable and encourage observers to complete their assigned observations. Do observers feel their supervisors believe in and support their efforts? Do supervisors facilitate the observation process by providing time and encouragement to observers?

Re-Administering the Perception Survey – Assess the impact on perceptions since the process began by getting all (or a sampling of) the workforce, managers and supervisors to retake the perception survey. Compare the new survey results to the first survey and see if desirable perceptions are more common.

Corrective or Continuous-Improvement Plans – For all deficits or opportunities identified in the assessment, formulate a plan to address them. Make the plan detailed as to actions and persons assigned to complete them, and set follow-up dates to ensure planned activities are accomplished. Keep the list of actions to be reviewed at the next audit. Plan a self-audit every year, or semi-annually, depending on how many issues need to be addressed.

Chapter Questions:

- Can the steering team audit their own process or will they need help at first?
- Are the major parts of the process functioning effectively and efficiently?
- Is the process adequately supported by managers and supervisors?
- Is the process impacting perceptions of safety?
- What corrective actions or improvements are needed?
- When will the next audit take place?

Conclusion

A lot of things that have been labeled as behavior-based safety simply are not. Many of the practices of mainstream BBS processes are ineffective and/or inefficient. The guidelines in this book are proven practices that have worked at many sites and in diverse cultures in both the US and other parts of the world. That said, every site is different and cookie-cutter approaches fail more often than they succeed. Ignoring cultural and workplace differences is the leading cause of BBS failures.

This means a good assessment of each site is a necessity and customization is critical. Knowledge of the site and the range of BBS options must come together in a strategy for implementation. The consultant leading the implementation begins the customization by selecting the best process strategies to fit the site's logistics. That strategy must also be flexible and allow for further customization by the site steering team, which will inevitably have deeper and more thorough knowledge of the site than the consultant or facilitator doing the assessment. These two steps of customization produce the implementation plan with the best chances of success. Even with such a plan, further customization may be necessary as the implementation rolls out and workplace realities impact it.

Customization must conform to the principles of Lean BBS, which are:

- Focus – and identify the significant few behaviors that most impact safety and get them into the heads and habits of workers.

- Influence – rather than confront. Remember people do things for a reason, and if you don't change the reason, you may not change the behavior.

- Listen – to find out what is influencing workers to do things the way they do and what gets in the way of desired changes in behavior.

- Measure – the key process indicators (KPIs) of your process to understand how activities impact behavior and how behavior impacts safety lagging indicators.

If you accurately assess, allow the steering team to customize, and remain open to lessons learned as the process progresses, your process will not only be effective, but will be sustainable long-term.

PS: The Future of Lean BBS

Describing Lean BBS is like shooting at a moving target. Every site implementation and every training of internal consultants yields new possibilities. Once you free yourself of pre-conceived notions of what BBS HAS to be, your imagination can really begin to explore. As long as you are true to basic concepts of what must happen, the ways in which you can make them happen are almost endless. A few of the most common innovations and improvements to the process we currently see among our clients are:

Even more narrow focus – Several sites only worked on one behavior at a time and solidly established that behavior into common practice and culture before moving on to the next. The cycle of these foci averages 90-120 days. The concept of "if you don't get it into their heads, you will never get it into their habits" is the first critical step. Adding other reinforcements besides just the observations is also common. Good sampling strategies for the observations is necessary to ensure that trends in the data are statistically significant.

Better management and nurturing of internal expertise – While ProAct recommends taking the internal consultant approach to ensure sustainability with internal expertise, we have not always been asked nor have we dictated exactly how many people should become subject-matter experts in Lean BBS. Many client organizations have developed a structure of support, from the corporate level to the regions to the sites, to ensure support and expertise from top to bottom. Many of these organizations have included this support function in their organizational charts and communicated it to everyone involved. Several have designated resources for ongoing professional development of these personnel.

Maturity models – After reading our STEPS book, many clients better realized the need for a stepped approach to safety improvement and the futility of trying to do too much at once. Based on this concept, they have developed maturity models describing practices in Lean BBS ranging from basic to advanced utilization. They encourage their sites to accurately identify where they are and periodically set improvement goals to move to the next steps of performance in specific areas.

Integration into organizational structure – Several safety departments have subdivided the behavioral and conditional aspects of safety and assigned personnel to each. Within each sub-department there are both mandatory and discretional issues, i.e. rules and regulations versus voluntary decisions. In such a structure, Lean BBS is the tool to address the discretionary behaviors as well as an information-gathering mechanism for all other aspects of safety. Very often, conditional issues can be influences on behaviors, and identifying them can enable corrective action from the best source. The conditional part of safety tends to be structured after the Lean BBS process as well. Conditional audits are performed similarly to behavioral observations in that they have a focus based on Pareto analysis and seek to identify exactly how the conditions could impact safety performance.

These are just a few of the ways in which the Lean BBS process is continuously improving. We are consistently pleased to see how the process not only produces improved safety performance, but innovative thinking among those involved. Best wishes on your journey toward safety excellence!

About the Authors

Shawn M. Galloway and **Terry L. Mathis** lead the global safety consultancy, ProAct Safety. They are authors of several bestselling books on Strategy, Leadership and Culture are two of the most prolific contributors in the industry authoring over a thousand articles, podcasts and videos. They have received awards and recognition by the American Society of Safety Engineers, National Safety Council and numerous magazines. Their consulting efforts have led over two thousand projects across all major industries on Safety Strategy, Leadership Safety Coaching, Safety Culture Transformation and Behavior-Based Safety.

Other Published Works

STEPS to Safety Culture Excellence

STEPS (Strategic Targets for Excellent Performance in Safety℠) demystifies the process of developing Safety Culture Excellence by breaking it down into small logical, internally led tasks. Although this book is dedicated to safety, the tested and proven STEPS process can be used to promote excellence in any aspect of operational performance.

Forecasting Tomorrow: The Future of Safety Excellence

Emerging trends are becoming increasing visible regarding how safety is viewed, thought of, strategically managed, and how progress is measured. Moreover, the cast of characters, the roles they play, and accountability for responsibilities are evolving, all for the better. The seven predictions found within this book are based on the authors' years of deep experience as trusted safety advisors across all industry sectors.

Inside Strategy: Value Creation From Within Your Organization

Inside Strategy is aimed at aligned continual performance improvement. Inside Strategy gives you - and everyone in your organization - a method of managing an unknown future to create new value for your internal and external customers, every day.

Index

[1] Galloway, S.M. (2011, January). Why We Fail to See Risk. *EHS Today*.
[2] Mathis, T.L. (2005, May). Lean Behavior-Based Safety: How the Process is Evolving to Survive in Today's Economy. *Occupational Hazards Magazine* (now *EHS Today*).
[3] Galloway, S.M. (2014, May). Who Should Perform Behavior-Based Safety Observations? *BIC Magazine*
[4] Mathis, T.L. & Galloway, S.M. (2010, February). Understanding Influences on Risks: A Four-Part Model. *EHS Today*.
[5] Mathis, T.L. (2009, October). Unions and Behavior-Based Safety: The 7 Deadly Sins. *EHS Today*.
[6] Galloway, S.M. (2013, September). Measuring Behavior-Based Safety: The Perfect Leading Indicator. *Occupational Health & Safety*

BBS Webinar Series

ProAct Safety leads a bi-monthly webinar series covering specific topics pertaining to BBS processes with the goal of making them more efficient and focused on delivering recognizable value. To ensure the sustainability of your unique BBS process, provide the leaders of your process and the observers that support it with ideas for continuous improvement and professional development. Annual subscriptions include access to all new and previous topics on-demand. For more information, visit www.ProActSafety.com/events.

Fully-Supported Implementation (Consultant led)

ProAct Safety's expert consultants can customize and implement the Lean BBS process and principles at your site(s) utilizing our proprietary, proven advanced techniques and training materials. The implementation process is unique to your operations and involves several visits over a one-year period and distance support between visits.

Public Lean BBS Workshop (Train-the-Trainer/Internal Consultant Approach)

ProAct Safety offers public Lean BBS workshops several times a year (as well as private workshops). Learn how to customize, implement, sustain and continuously improve your own BBS process. Utilize advanced techniques from the most recent breakthroughs in the industry, and become an internal consultant and subject matter expert for your company. This approach is ideal for companies who want to maximize their ability to internally implement while minimizing external consultant costs.

For more information on options, how to customize the approach for your culture and operational realities, and upcoming dates for public events, visit www.LeanBBS.com.

Made in the USA
Columbia, SC
03 March 2020